Storing Your Home Grown
Fruit and Vegetables

Storing Your Home Grown
Fruit and Vegetables

HOW TO MAKE YOUR GARDEN'S BOUNTY LAST ALL YEAR ROUND

PAUL PEACOCK

howtobooks

Published by Spring Hill

Spring Hill is an imprint of How To Books Ltd
Spring Hill House, Spring Hill Road,
Begbroke, Oxford, OX5 1RX, United Kingdom
Tel: (01865) 375794 Fax: (01865) 379162
info@howtobooks.co.uk
www.howtobooks.co.uk

How To Books greatly reduce the carbon footprint of their books by sourcing their
typesetting and printing in the UK.

First published 2011

British Library Cataloguing in Publication Data
A catalogue record for this book is available from the British Library

ISBN: 978 1 905862 54 2

Text illustrations by Rebecca Peacock
Produced for How To Books by Deer Park Productions, Tavistock
Typeset by TW Typesetting, Plymouth, Devon
Printed and bound in Great Britain by Bell & Bain Ltd, Glasgow

NOTE: The material contained in this book is set out in good faith for general
guidance and no liability can be accepted for loss or expense incurred as a result of
relying in particular circumstances on statements made in this book. Laws and
regulations may be complex and liable to change, and readers should check the
current position with the relevant authorities before making personal arrangements.

CONTENTS

6 A–Z of Growing, Storing and Preserving Vegetables   **78**

Growing the best veg doesn't stop until it is properly preserved

INTRODUCTION

Growing your own

It is a rite of passage. You decide to grow some of your own food. An impulse frequently charged by the world of supermarkets disgusting you so much that you decide to live a cleaner, less tainted sort of life, in which the four winds bring sunshine and rain to your plot and Mother Nature smiles beautifully on your crops, and you, in your sleeveless V-necked jumper sporting a beaming smile, wheelbarrow home your produce that has been seasoned with blue skies and high sunshine.

But you quickly learn about weeds, and insects that nibble at your crops, and the unhurried evil of slugs and snails, and fungal infections galore; and quite rightly. Nature, who wonders in variety, will use your crops to feed so many other living beings that, if you are not careful, there will be none for you and your family.

So, preserving your hard-grown crops is a matter for absolute attention if you are going to make what you have last for any length of time.

The fundamentals of this book

There are two basic ways to think about self-sufficiency. One is that you grow crops to last you through the year and the operation is about getting enough food to feed yourself in the time and space limitations you have. Another way of thinking about self-sufficiency is to produce the very best food and try to make it last for as long as possible.

This book is about the latter: growing the best food and trying to make it the best food for the longest possible time. And there is a good reason for it too – the UK is particularly blessed with mostly wonderful soil, plenty of rain and a very reasonable climate. We should be able to grow wonderful produce!

But what happens when the food runs out? Well, you either have to buy or starve. But it is possible, even in a small garden, to produce enough food to last the year through, or with a very short gap until your next crop is ready for harvest.

So this book looks at a number of things. Obviously there are recipes for jam and pickles and other ways of preserving. But there is also information on how to extend the growing season so that you have fresh produce in the ground for longer waiting to be picked.

A major feature of this book is how to grow food so that it keeps best. So it offers a number of techniques that ensure the very best storability of crops, as often the techniques needed to store a crop are different to those needed to create fresh produce.

Emphasis on the best

I'd rather have a ton of really great potatoes than a ton and a half of poorer ones, and that is the basis on which this book approaches the art of preserving food. The best-grown food also keeps for the longest. Sometimes there is a need to grow food that we are going to preserve in a different way than the food we are going to eat fresh, and that is mentioned where necessary.

We look at ways of growing crops and organising your garden, so you get enough food when you need it or, better still, enough to store in some perfect way to give you safe and tasty food all the year through.

The heart of being self-sufficient is sufficiency, as far as possible at the right time. In other words, you might have excellent food just when you need it, and a little on the side while you are waiting. Yes, it is possible to have tomatoes in August and September but it is nice to have them in December too, if possible. You don't always want to have a bottled tomato – sometimes you need a tomato. Now, of course, you can go to the supermarket! But supermarket tomatoes will have come from another continent, and that's not self-sufficient.

Many people find gardening a chore and hard work and find it difficult to maintain the zeal they started with right through the season. Maybe the promise of some marvellous produce and what you can do with it will keep you going. And so this book concentrates on the following ideas:

- Increasing the cropping time for plants, so that you get a good crop over an extended period.

- Growing food so that it is at its peak for eating and for storing (often two different things altogether).

- The best ways to store fruit and vegetables intact, extending their shelf life as far as possible.

- The best and most delicious ways to preserve fruit and vegetables.

What about freezing?

There are some who feel that freezing is anathema and shouldn't be used because it wastes energy, contributes to global warming and therefore is a poor way to store food. In a way I agree but, that said, if I had the ability to generate my own green energy, I would use a lot of it to run a number of freezers. Freezing is as legitimate as preserving with salt and sugar. In this day and age, the ability to buy a bag of salt from the supermarket belies the fact that this salt costs the earth in energy use, transportation, processing and packaging and has a carbon dioxide cost too.

So yes, there is a section on freezing in this book. After all, we all have freezers! But if this were the only way we preserve our food it would be a boring cuisine we have. We want to celebrate pies, tarts, jams, preserves, pickles, juices, wines, ciders and a myriad of delicacies that simply wouldn't exist if all we did was freeze.

However, this is not the sole point of this book. Naturally, we have been keeping fruit and vegetables for many hundreds of generations and many of our best foods and ingredients have been stored in some way – the jams and preserves, the salted and the dried, indeed, almost all the wonderful British produce many of us pick from supermarket shelves.

The majority of our harvests can be preserved in the natural way, and that is what this book celebrates: techniques, recipes and good growing.

CHAPTER 1

EXTENDING THE GROWING SEASON

This chapter is based on the general principle that gardeners need to produce crops over an extended period, and that in many cases fresh is better than preserved.

This book aims to make sure that you have good, wholesome, safe, tasty and edible food for as long as possible around the year.

Most of us have become accustomed to eating fresh food any time we want it, bought from the supermarket. You can buy cabbages from every continent, shipped or flown into the country almost on demand. However, this means that we have lost some of the seasonality of food, and are not used to the idea that certain kinds of produce are available only in certain months, relying for the rest of the year on bottled, tinned, cooked or frozen versions of our favourite foods.

The vast array of tinned foods people buy at the supermarket because they are either convenient, easy to store in the kitchen or just lovely to eat are there because at one time this is how people stored their food and they have now become ingredients in their own right. For example, pickled onions are not like onions, but we love them all the same. They contain just enough vinegar to make them tangy and tasty, and just enough sulphur to remind everyone about it the following day. But a pickled onion is pickled – it's not an onion.

This chapter is about extending the growing season so that you have, and unfortunately onions are a bad example, proper, real, un-messed-about-with fresh vegetables for as long as possible throughout the year.

GETTING FRESH FOOD FOR AS LONG AS POSSIBLE

It is fair to say that plants evolved to be in the ground, growing. They, like us, have a highly evolved immune system which keeps them fresh while they are growing. Eventually, however, the changing of the seasons makes it increasingly difficult for plants to stay in a fit condition and then they have to go into storage. But for many plants there is a window of opportunity in which it is possible to have them growing in the ground, to be harvested over time.

It is a simple idea and one which provides the fundamental basis of self sufficiency: crops are best in the ground whenever you can manage it.

SUCCESSIONAL PLANTING

In many cases there is, more or less, a fixed length of time from sowing a seed or planting a crop to it being ready for harvest. If you sow a lettuce plant in April, you might get a crop in June, but there would be still time enough left in the year to get a crop if you plant in June and have a harvest in September. So, for most intents, you can sow from April to June, and get a pickable lettuce from June to September. This is successional sowing or planting.

A quick word about quality

There are a number of aspects to be aware of. First, you should store the very best quality produce, and eat the rest. The major storing crops come right in the middle of the season. Those forced to be either early or late should be kept for more or less immediate eating.

The second point is you are not growing crops for sale in the supermarket and some of them – especially those grown completely out of their season (like potatoes for Christmas) – are prone to look second class. Does this really matter? Vegetable and fruit gardening is about quality, of course, but at the same time, should we discard something just because it doesn't look right or because we have to trim it a little in order to be able to eat it?

Seeds

Successional sowing is more frequently a feature of produce grown from seeds than potatoes, carrots and turnips. Sowing over many weeks to give a longer harvest is the norm for almost all your salad crops.

As a general rule, the shorter the cropping period, the more useful this method can be. Long-growing crops like parsnips and leeks tend to catch up with each other to mature more or less at the same time. All the onion family – onions, leeks, shallots, chives and garlic – respond to the length of the day in order to set their metabolism for the following year's growth. Consequently there is not much point in successional sowing of alliums, as they are called.

Carrots, however, do very well sown every fortnight from April to August, giving you fresh carrots in the ground from June to October. Yes, the June ones are small and the October ones are prone to frost problems, but you do have fresh carrots.

Similarly, I find the variety of cabbage All Year Round does well with successional sowings, which I do on a fortnightly basis in each of the four seasons. You can have cabbages in the ground every day of the year so long as you protect the winter ones with a cloche.

EXTENDING THE CROPPING SEASON

Let us take potatoes as an example of extending the cropping season. Different varieties of potato take longer than others to come to maturity. What we refer to as first earlies can take 12 weeks to come to harvest, second earlies can take 16 weeks and those known as maincrop can take 20 weeks.

So the variety Pink Fir Apple is planted on 17 March (the traditional day for first earlies) and is ready for harvest in the middle of June. But there is no reason to lift your potatoes all on the same day. Indeed, you can leave them in the ground for a month beyond this. So you now have what we call new potatoes ready for eating, fresh as you like, from the middle of June to the end of July. Furthermore, King Edwards can be planted in early April and might not be ready until the first or second week in September.

So by planting different varieties you can have a ready supply of potatoes fresh from the ground from June to the end of September. But there is a problem: these potatoes are quite different in their cooking characteristics. You wouldn't think of making chips with Pink Fir Apple for example – they simply have the wrong starch content. This is the reason why there are so many potato varieties around.

EXTENDING THE GROWING SEASON UNDER PLASTIC, IN A POLYTUNNEL OR GREENHOUSE

It is possible to create a suitable environment for growing. Potatoes, for example, burst into life by a number of mechanisms, but mostly due to temperature. When the soil temperature reaches an appropriate level, the buds on the tuber burst into life, giving a new plant. There are a number of tricks for creating adequate soil temperatures to make many plants start off earlier than they otherwise would have done. You can plant in a polytunnel or greenhouse, use plastic cloches or even plant in tubs in the house and take them outside later in the spring. This means that it is possible to grow, for example, potatoes, carrots, turnips and cabbages for a brilliant Christmas dinner.

Creating the right environment for growing works for many different plants but mostly root crops. It does not work so well for plants where the length of the day is the trigger for germination or growth. It can be done, but you need day-glow bulbs to boost the day length, and wiring a greenhouse is not all that convenient.

You can have potatoes for Christmas by planting first earlies in tubs in August and moving them into the polytunnel in September. You can have carrots, turnips, salads and many herbs all the year round.

You can add at least a month to the early side and two months afterward to the fruiting/cropping season by growing under cover. By partitioning off a large polytunnel you can have seeds germinating at 14°C, ready to put into their growing positions at least a month early, and successional sowings will extend this further.

My experience has been that growing under cover is expensive in terms of space – you can't grow everything in there! But you can grow enough to enjoy early crops that you simply otherwise wouldn't get.

The black plastic trick
In the middle of February, cover the ground with black plastic to heat the soil using the weak late winter sun, and place a cloche over the top to warm the ground further.

Some people I know use a layer of black plastic, followed by a layer of clear. This makes the inch or so of air above the black plastic heat up. The soil beneath is several degrees warmer than bare soil, and if you plant in this your potatoes are earlier by a week or more.

Using cloches, cold frames and hot beds

Cloches will speed the growth of some plants, especially fruit like strawberries, salad produce, carrots and young onions. Both the cloche and the cold frame allow plants to grow unhindered by the wind and rain, therefore producing more perfect plants, which will consequently be better for storage.

We tend to forget that the cold frame can be used to grow plants to maturity as well as being a place for hardening off. Cucumbers grown early in cold frames are super.

Hot beds are best dug in greenhouses. Simply dig a hole that is at least 75 cm deep. (Use all the soil you dig out elsewhere in the garden.) Then three-quarters fill the hole with fresh farmyard manure and top it off with fresh compost. As the manure rots, it generates heat which will keep the compost warm. Give the bed a month for the heat production to even out, then use it either to sow directly or to stand pots for a heat blast.

EXTENDING THE GROWING SEASON OF FRUIT

Extending the growing season of fruit isn't so easy. Fruit bushes, like blackcurrants and raspberries, fruit in their season and that's that. You cannot do much about it other than to grow various varieties that crop early, mid season and late, giving you different varieties at different times of the year.

The following information is offered solely for choosing varieties to extend the growing season. More advice on growing and preserving is included elsewhere in this book.

Apples

There are still hundreds of apple varieties – even though many have been lost in recent years. The apple is traditionally an autumn fruit, the majority of plants coming into season in late September. But by judicial variety choice, you can have apples in season from September to November.

- Irish Peach is usually ready in late August, but it can be earlier depending on climate

There are so many apple varieties, you are advised to make a search of the nurseries, rather than look for an exhaustive list. Anyone wanting to grow apples for as long a supply as possible should look around and plan to have fresh apples from August to November, and add a four-month keeping cycle of fruit after that. In essence, you can have apples from August to about April.

Blackcurrants

You can have fresh blackcurrants from July to August by having bushes of different varieties, though you can enjoy the leaves for tea from much earlier in the spring.

- Ben Hope sets fruit in the middle of June and is full by the end of June

- Ben Gairn sets fruit in the last week in June and is full by the second week in July

- Ben Tirran sets fruit at the end of July and is full by the middle of August

Gooseberries

If you grow the variety Pax out of strong wind and the variety Invicta against a wall, as a cordon for example, you will have gooseberries from May until August.

Raspberries

There are lots of varieties from Scotland that will allow you to grow decent fruit quite early. They come in two main types, summer fruiting and autumn fruiting, giving a really long supply.

SUMMER FRUITING

- Glen Moy is a mid-season raspberry in Scotland, but in England it will produce fruit in June

- Glen Ample will produce fruit in mid July and peak at the end of July

AUTUMN FRUITING

- Autumn Bliss fruits until late August

- Autumn Treasure will stop fruiting in early September

COLLECTING BRAMBLES AND WILD RASPBERRIES

The countryside and town are awash with brambles from July onwards, and they are clearly available to those who are happy to pick them. Choose a supply you know have not been urinated on and, while we are at it, don't pick them from late September – the Devil is supposed to pee on them!

You will not find them to be the best quality, and they are certainly not good for jam, but they do make excellent wine.

Strawberries

Strawberries can be coaxed into setting fruit early by covering them with cloches from winter onwards. This way you should have fruit from late May.

Plants like strawberries have a peak period when the fruit is both plentiful and of best quality. Fruit for preserving should always be taken from this peak time.

- Christine can have cloched fruit from late May and peaks in late June

- Honeoye starts to fruit in mid June and peaks in early July

- Cambridge Favourite fruits in late June and peaks in mid July

- Florence fruits at the end of June and peaks towards the end of July

- Malling Pearl can bear fruit from June to September and is known as an 'ever bearer'

Consequently it is possible to have strawberries from the last week in May to September, but the best fruit for preserving are to be found in late June – early July. However, if you need to there is nothing to stop you making jam in August and June too!

EXTENDING THE GROWING SEASON OF VEGETABLES

Aubergine

Sow them in February and grow indoor and outdoor types. The small varieties, such as Baby Belle, produce earlier fruit.

Beans (runner, French etc.)

You can have a harvest that lasts the whole summer long if you follow the rules of constant watering, daily picking of fruit and feeding once a week. For each fruit picked you get another flower to replace it, and it is possible to have beans from June/July to September.

Beetroot

Normally sown directly in the soil, try some in paper pots made from newspaper. You can sow these indoors at 15°C and get them going in late March. Keep them cool once they have germinated, and then in May peel the paper away and plant them into holes made with a dibber. You need to be really careful when transplanting, and you only get small fruits, but they are lovely and sweet.

They also grow pretty much all the year round in a polytunnel.

Broad beans

This is one of those plants that can be over wintered. They are quite frost hardy and sowing in September gives little plants that stop growing once it is cold. Then, in the Spring, as soon as the weather warms up, they shoot up and you get fruits that are about a month earlier than the April-sown plants.

Cabbage

It is true to say that you can have a cabbage almost every day of the year if you plan accordingly. But beware: the sulphorous compounds in the cabbage will play havoc with your personal exhaust system, so be prepared to spend a good few hours standing outside when you grow cabbages in the winter.

You can sow cabbage directly in the soil, even in early spring if you protect it with a cloche. However, it is frequently best to sow indoors in large modules or 8 cm pots. Use lime and sow about 3 seeds per module/pot. Discard all but the fastest growing plant. If you sow in March, by May you will have large seedlings to plant out.

You can quite literally sow the variety All Year Round all the year round, so long as you have reasonably warm temperatures for germination.

Carrots

You can prolong the growing season by growing more than one variety. Durham Early will crop a month earlier than the others. Also, you can sow a row of carrots every two weeks from April to August. Those sown in August can be covered with a cloche in October. This way you get carrots for most of the year. They keep well too – so there is never any need to be without them.

Carrots do not transplant very well, so there is nothing to be gained from sowing indoors.

Cauliflower

You can sow every two weeks from April until late May – 3 or 4 sowings – which will stagger the harvest, but they do have a tendency to catch up with each other.

Courgette

These plants often do not respond to early sowing, and tend to come into fruit at the height of the summer.

However, you can increase the cropping period by constant cropping. Take courgettes when they are about 15 cm long and you will always get another flower to replace it. This stops in September, so you get a good summer's worth.

Garlic

All the books say you need to plant garlic in the winter, but the truth is that this plant is useful planted at any time so long as the soil is warmish. Sure, corms planted in May will never really produce a large crop, but you will get one.

Secondly (and I shudder at my own advice), the books say only grow dedicated varieties for the UK from a proper supplier. However, I use supermarket garlic planted anywhere in the garden as a companion plant and generally they produce usable corms at any time of the year.

Peas

You can extend the pea season from June to September/October. First of all you can sow successively from February to June. I find Kelvedon Wonder to be a brilliant all-round pea.

Secondly, you can grow indoors in the tunnel and then outdoors at the same time giving you an earlier crop.

Potato

We have discussed this plant above. It is possible to have fresh potatoes from June to September, and they keep for ages afterwards too.

- First earlies: Maris Peer, Home Guard, Arran Pilot, Pentland Javelin, Rocket, Pink Fir Apple
- Second earlies: Kestrel, Wilja, Estima, Osprey, Nadine
- Maincrop: Admiral, Cara, Eden, Maris Piper, King Edward and all those blight-resistant varieties listed in Chapter 3

Rhubarb

The way to extend the rhubarb season is to force them in the spring. As the plant appears after the winter rest, place a dark pot over the emerging stalks – exclude the light completely. Inside the stems will etiolate (which literally means 'reach for the stars') and you will get early rhubarb.

Salads

So long as it is warm, you can grow lettuce, radish, beets and other salad leaves. None of these plants like to be transplanted, so don't bother trying to sow in pots indoors to be transplanted later into the garden. A sowing every two weeks and indoors from October onwards, will provide you with a good all-year-round supply.

Tomato

The cropping period for tomatoes basically lasts from August to late September. You can increase this by sowing early under heat and hardening off the plants and growing some outdoors. This is particularly successful with the variety Gardener's Delight which seem to do well out in the garden. Of course, the next problem is ripening, which we will look at later.

CHAPTER 2

HOW TO HARVEST

This chapter shows you how to harvest your produce without destroying all your careful work in growing it in the first place.

Anyone that spends time with older people will pick up something of a respect for food. There are lots of reasons why we need a healthy respect for food. It is not over melodramatic to say that you are what you eat, that food should never be wasted because it is precious and that it should always be handled with respect.

In terms of harvesting, your handling of food is very important. It is an exciting time, and we often allow that excitement to get the better of us. It is important that plants are harvested both with care and diligence because this is going to be your food for the next year. Harsh handling, over zealousness with fork, spade or knife and incorrect treatment once the plants are collected severely reduces storage qualities along with nutritional value.

FIRST FRUITS AND THE LAND/ CROP CYCLE

If you think about it, keeping food in storage is not the only goal of the gardener. Harvest is such a great time, a time of celebration, so make sure you do celebrate with the first fruits of your labour. Nothing, for example, beats new potatoes, harvested and cooked within a few minutes. I, for one, couldn't forego such pleasures.

Before you collect your harvest, try out your crops. This is not only a celebration but a chance to assess the quality of your crop, and therefore how you might choose to preserve it – if at all.

You also get an idea, over time, how your land or garden responds to certain crops, which part of the garden is too wet for potatoes or too stony for carrots, for example. Comparing, year on year, how crops do is the best guide to land improvement.

We ought to have perfect soil in this country. In a country where there were 5 million allotments during the war, we should have a continuum of one generation following another, each improving the land. Alas, we left the soil for the convenience of the supermarket, and now we have to learn again how the harvest informs the next few years of growing and work in the garden.

USE A TRUG

A wooden trug is a shallow basket where produce is placed, and I would like to make a plea for its reintroduction into the garden.

It has to be said: there is an elegance in a trug. Full of produce, they look the part, and this is the best advert for good design – looking right. A trug, if you can manage to get your hands on one (and they are not cheap), protects the harvested vegetables from being bashed around. The gently sloping bottom will let the produce roll but not crash.

I also like the open nature of a trug: it lets the vegetables evaporate water without causing condensation. Of course, from the second you harvest a plant, it starts to deteriorate. Loss of water in something like a plastic bag can cause infections in plants. Many's the time I have put potatoes into a supermarket bag only to find them rotting a week later when I have forgotten them.

The use of a trug is reflected in how you consequently handle the harvest once collected. I once dragged a large bath I found on an overgrown plot on one of the allotments (isn't it a terrible thing, an overgrown allotment?). Into it, I threw my potato harvest as I lifted it from the heavy soil. Within a week it was showing signs of rotting. The merry ting as the spuds hit the sides of the bath haunted me, with every collision between potato and enamel there was an associated bit of damage to the crop.

BE GENTLE

Whether you use an Old English wooden trug or the wheelbarrow, be gentle with your crops. Place them down rather than throwing. Order your harvest with spuds at the bottom and lighter, more delicate crops on the top. All these obvious things will make for a splendid harvest.

QUICK WITH LEAVES

As we have said, harvested plants deteriorate quickly. Leafy vegetables, especially, evaporate water very quickly, and the texture and flavour of the food change dramatically. Spinach, for example, will not rest many hours before becoming inedible.

So don't hang about before the next stage; you are looking to make the best meals from your crops, either straight away or sometime in the future after storage.

Leaves (cabbages, spinach, lettuce, beet, etc.) are there to evaporate – it is part of how plants live. Once removed from the soil, from their water supply, they wilt and change quickly. Leaving them in the ground is the best way of keeping them fresh. Whether they are destined to be cooked that day, or whether they are going to be turned into some preserve or other, I always plunge them in cold water as soon as they are harvested. This cools and moistens the plant, reducing the impact of evaporation.

I tend to grow flat-leaved lettuces in individual pots on the patio, allowing me to bring the whole plant into the kitchen, pot and all, avoiding floppy leaves. Alternatively, you can cut and come again, that is grow in pots near the kitchen door enabling you to take just a couple of leaves and then leave the plant to grow some more before taking from it again.

HARVESTING POTATOES

Lifting, not digging

We 'lift' potatoes and other root, tuber and rhizome crops, and the term gives you a clue about the process. Sure, in a 30 acre field which yields 100 tonnes of potatoes, they are treated harshly, lifted by a potato plough. But in the garden we can be much more gentle than that.

First of all, remove the vines as far as you can, leaving only a stump. This is to be sure you have no blight, or other fungal infections, affecting the harvest from the leaves, and it also gives the soil a chance to dry out a little (if the weather allows). By the way, I never compost potato vines because they might well have fungal spores that you will then pass around the garden the following year. Instead, I burn them and then use the ash on the garden.

Use a garden fork to loosen the soil a long way from the centre of the plant. On lifting a fork full, you will dislodge the soil and some potatoes with it. Repeat this all around the plant and then get your hands in. Lift potatoes individually, by hand.

Wipe any soil you can from the potato and then leave those for storage to cure.

Be sure to get all the tubers up including the tiny ones – they will grow next year, but may well have fallen foul of viral infections. Always buy new seed potatoes each year.

Curing potatoes

Potatoes destined for storage should be cured. This is done by simply leaving them to air (keep them dry) and the skin will harden and become impervious to further evaporation, and also infection from outside. This takes about 3–5 days, and thankfully, when you are lifting your potatoes, the weather is usually dryish.

LIFTING CARROTS, TURNIPS, PARSNIPS, SWEDES

Use a garden trowel or a small hand fork to lift the root. Don't rely on just pulling it out by the leaves; they invariably pull off. You can grab the leaves and lift with the trowel from underneath.

Once harvested, remove the leaves to reduce evaporation from the root. Have a feel of your root – it should be hard, not floppy. You are looking to keep them hard in storage. Floppy roots are difficult to cook, and are a direct consequence of evaporation.

Root crops for storage should be cured like potatoes before they are either shelved or more traditionally, and more successfully, clamped (see Chapter 4).

ONIONS, GARLIC, LEEKS

To be honest, I never have tried to store leeks. There is little point. They sit in the ground, sometimes sadly and forlorn, alone and neglected, but once pulled are ready for eating straight away. Also, they cope well with frost.

In order to help them last the longest time possible, trim the leaves in the winter and cover with fleece.

Onions will tell you when they are ready for harvest: their stems will fall over and the bulbs will be firm and cricket-ball sized.

Do not try to pull them from the ground by their leaves. Lever them out with a trowel and trim both the leaves and roots. Cure them like potatoes – the outer leaves will become brown, leathery and dry, protecting the flesh within. Always take them away to cure; don't leave them where they were growing because the aroma of onion attracts pests.

Japanese onions, grown from late August and over wintered do not keep so well as spring-sown onions. Always treat Japanese onions with 'kid gloves' – don't bang them about during harvest because they will bruise easily and keep even less well than normal.

Garlic should have their leaves and roots trimmed and stored as whole bulbs, leaving the individual corms inside. They too should have their roots and leaves trimmed to cut down evaporation.

BEANS, PEAS, RUNNER BEANS, MANGETOUT

If you are harvesting beans or peas for storage, by either drying, salting or baking, remove the seeds as soon as possible after harvest and get on with the process straight away. Otherwise the outer seed cover will be tough.

If you are salting runner beans, trim the spines and the ends straight away after harvesting them. Take these beans as early as you can in the morning – they keep better then. If you wait for the end of the day they are never at their best as they have dried too much.

Use only the just ripe bean pods for this process, leaving older beans for immediate consumption.

Peas should be shelled straight away, on the same day as pulling, and then processed.

CAULIFLOWERS

If you are thinking of lifting cauliflowers for freezing, collect just ripe, white plants. Do not bother with them once they are going over to cream. In effect you are taking only the very newest flowers.

Plunge them in a bucket of cold water straight away after picking, to cool them down and keep them fresh until you get them to the kitchen. Don't forget to return to the garden to remove the roots.

TOMATOES, CUCUMBERS, PEPPERS

Ripening is an awkward subject with tomatoes. Some place a ripe banana near the un-ripened fruit so the ethylene gas, which is a plant hormone produced by the ripe banana, works on the tomatoes and ripens them too. However, tomatoes ripened in this way do not taste as good as naturally ripened ones and the ripe banana can also introduce fungal infections that cause the fruit to store less well.

Tomatoes for processing should be perfect and fully ripe. Only the most perfect fruits possible should be used to make green tomato chutney, even though they will be the dregs of the crop at the wrong time of the year.

Cucumbers can be collected as soon as they attain a decent size, and if they are to be pickled, plunge them into cold water on collection. Gherkins are treated in just the same way.

Peppers can be left on the plant until you are ready to gather them as a whole crop.

HARVESTING FRUITS

Apples

Apples are ripe when they just fall from the branch. Collect them in the morning for best storage and wrap them in tissue. Only store perfect apples, and I mean perfect.

If you want to juice your remaining apples, do it straight away if you can. You will then have to pasteurise the juice, which should also be done straight away (see Chapter 4).

Soft fruit

For best results, keep strawberries on straw avoiding the grit and rain splashes, and most of the slugs.

Take all soft fruits from behind the fruit – don't hold the berry if you can help it as you will damage it. As you are collecting the fruits, keep them covered with a damp towel, especially if it is a very hot day, to keep them in their best condition.

Prepare them for storage straight away. If you are making jam, try to do it on the same day they were picked.

Gooseberries

These are so full of sugar and so prone to fungal attack, it is best to take them as soon as they are ripe. Try not to leave them on the bush. Pruning is also important – make an open, cup-shaped plant so the wind can blow through.

Grapes

Grapes can be used for jam, jellies and, of course, wine. Keep an eye on your grapes; don't let the bunches fill out so much that there is no space between the berries – use scissors to remove the odd berry if necessary. Collect whole bunches and process immediately. Plunge in cold water on collection and only ever process perfect berries.

CHAPTER 3

STOPPING YOUR HARVEST FROM SPOILING

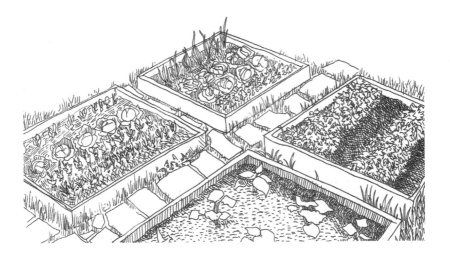

This chapter looks at how food is spoiled and in particular how we can, while we are growing it, keep it in tip top condition for storage.

The spoilage of food is an amazing proof of the absolute abundance of life on our planet. Life bursts into every tiny crevice of food, each individual organism taking supreme advantage of what is available, and there is nothing you can do about it.

Around 300 years ago people saw this as the creative God producing life from nothing and the invention of the microscope provided actual evidence of this at work. A whole class of microscopic animals, then called infurosa, appeared wherever food was left in water. We now know, of course, that these animals were to be found in water anyway, and the presence of the food just made them reproduce more quickly. The fact is that what the scientists of the day called spontaneous generation was simply the living world taking advantage of food.

WHY FOOD SPOILS

Microbial action

Food is spoiled by fungi and bacteria, each devouring it to produce more of their own kind – they essentially live in the food. Usually, fungal infections smell and produce a furry kind of fruiting body. Bacterial infections cause food to discolour, bruise and frequently turn mushy as it is devoured from within. On the whole, these two spoilers are using the energy in the sugars to keep themselves alive. They can grow very quickly and spoilage can transfer from one infected food item to the next.

Oxidation

Food often spoils because it oxidises in the air. This is difficult to deal with because there is a lot of oxygen out there. Sometimes, burying roots in the ground keeps them just perfect in this respect. Wrapping food in tissue helps, as does vacuum sealing.

Oxidation also occurs at cut surfaces, not just because of the air but by the spilled contents of cells that exude enzymes into the food. This also happens if food (such as apples) is dropped.

Dehydration

Putting food in a sealed package or in a place where constant evaporation is not possible keeps the moisture in place and the contents in tip-top condition. Often, humidity is the promoter of disease, especially fungal infections. If you can keep this to a minimum, you will severely reduce the spoilage of the crop. This is one of the fundamental principles behind clamps and storage pits, which are used to create just the right conditions for keeping the plant fresh.

Of course, you cannot always store produce in this way, and a grape will resemble a currant if left for some time. Keeping delicate fruit in cooled boiled water was often used to prolong its freshness for many days. There are many other examples of produce being kept for a long time by avoiding dehydration. Toffee apples, for example, actually last a long time, as do waxed fruit of all kinds, not seen these days because of modern food processing. The use of soda glass, sodium silicate dissolved in water, was used not only to keep eggs fresh but to paint many fruits and roots that were later to be peeled prior to cooking.

Contamination

Food can be spoiled by contaminants, either from other food or by anything in the journey from plot to plate. Bad water and poor storage conditions are frequently to blame, but animal action is also a problem. I once had a large consignment of potatoes ruined by rats peeing and defecating on them. They had not been stored in sacks, nor removed to a rat-proof area.

Animals

We have to remember that spoilage can also be caused by insects, insect grubs, rats, mice, family pets (we had a cat who thought a sack of potatoes made a great litter tray) and even children. It is important that food is stored in such a way that these agents cannot get at it. Always store produce in a secure way. Lockable steel boxes or tins are frequently used – I buy up old army surplus ammunition cases, but an old lockable three-tier filing cabinet is just as good.

Physical factors

Food that is roughly treated is easily bruised, rendering it quite useless for keeping, though you might not know it at the time. Many a crop of apples, and other fruits, have been ruined in this way.

You drop an apple, store it in a sack with other apples, and the dropped one very quickly becomes rotten, infecting the whole sack. The impact damages and breaks cells inside the apple. The cell contents spill out into the tissue and usually the action of enzymes breaks more cells. Now, inside the fruit, there are broken cell walls, sugars, oxidising enzymes breaking more cell walls – a perfect environment for certain fungi to take advantage. (By the way, the fungus that most often takes hold in these circumstances is penicillin but there is no use trying to eat it, you'll only get sick.)

AVOIDING SPOILAGE

All of these factors that potentially spoil food can be countered in a number of ways.

The plant's own protection

Plants do have their own measure of protection from diseases. Working with this is a matter of firstly not allowing too many plants to grow in one place and only storing foodstuffs that as far as you know are in perfect condition.

The drying effect of the air and storing food with space around it is often enough to keep food sufficiently well for use. This applies particularly to root crops.

Curing

This method relies on the plant's own immunity to resist infection. It is not surprising (to anyone who has peeled swede or turnip) that root crops have quite a resistance to infection on account of their thick skin, as do onions because of their leathery outer leaves and strong sap.

Cooking

Raising the temperature kills everything in the food, and thus it will keep for at least a week in a clean, sealed container. Obviously, if the container is air tight and sterile before use, you can keep food for quite a long time in this way.

Osmotic pressure

If you keep food in a strong salt or sugar solution, the osmotic pressure of the chemicals draws water out of the bacteria or fungi, stopping spoilage in its tracks. Osmosis works by the affinity that charged particles have for water, and water will burst out of bacterial cells and fungal hyphae, killing them.

Poisoning spoiling microbes

Keeping food in vinegar actually poisons bacteria and fungi. They simply cannot live in the acidic environment.

Freezing

This doesn't kill spoiling microbes, it just slows them down so much that they cannot reproduce. They frequently speed up their reproduction once the food is thawed, often at a greater rate than before, so frozen food does not keep so well after freezing, and it is certainly unsafe to refreeze thawed food.

Combinations of methods

Most storage methods in fact use more than one method – the belt and braces approach. Pickling usually involves three: cooking to kill germs, salt or sugar as an osmotic precaution, and vinegar as a poison to any microbes that might fall in the jar when the lid is opened.

Maintaining optimum storability

From the moment a seed germinates, it is attacked by living things, physical factors and the harshness of the environment. The growing plant makes a concerted effort to counter all the forces that are against it and builds up its immunity, usually in the form of thick coatings, resins within the plant and in some cases special antibacterial, antifungal chemicals.

In order to optimise this natural storability we need to treat growing plants correctly, in terms of watering, nutrition and handling. A wrong move at an early stage of the plant's development can result in produce that does not keep as well as it should.

GET THE GROWER'S MINDSET

The unfortunate thing about the passage of time is the way our attitudes change. I don't just mean individuals, but whole generations. Once at the forefront of our way of life, the garden has been replaced by the street, the village by the town, the town by the city. In times past, everyone had their food garden and most of the family's food came from it. But more than this, since the garden was often the family's main creation, it was cared for in a special way.

The care gardeners had for their crop changed with industrialisation. Trugs gave way to hoppers, hand picking became machine harvesting, and this has had quite an effect on the way we now handle our food. Old gardeners cut, caressed, coaxed and all but lived with their crop. These days we plant, water, feed and pick. There is a harshness in the way we garden – and I want to find again that relationship between gardener and his or her growing food.

The way we treat our plants will determine their final state, sitting on a shelf in a store cupboard. You might not notice that the dropped potato is damaged inside so that it will eventually spoil more quickly than the others, or know that holding a seedling by the stem might allow a fungal invasion of the plant that, if it doesn't kill it there and then, will lead to ruined crop later.

And, of course, pests and diseases that hit our crops while they are growing have significant implications for the crop later on.

Handling plants

This is so important and it starts with seedlings. If you can help it, do not handle seedlings. Grow them in modules, leaving you with one plant per module ready for potting on to their final positions.

If you must handle seedlings, always do so by the leaf, as there is little further consequence if it is damaged. If you hold the stem, however delicately you do so, you will damage some of the tissues inside, leading either to death straight away, or to storage problems later.

Pruning

Walking around the garden, pruning fruit or controlling growth, restricting fruiting tomatoes and a hundred other jobs, make sure you clean your utensils between plants. I keep a bucket of washing soda in water – about a handful per gallon – to clean the secateurs between tending to plants. This way I avoid transferring problems from plant to plant.

Stop walking on the soil

This is probably the worst thing anyone could do in the garden. Walk a plank! Especially on allotments, people spread spores all over the place by simply walking on the soil. This is the reason why there is hardly an allotment in the land without a club root problem.

Disinfect

Make sure your greenhouses, old pots, spades, polytunnels – everything in the garden – is given a good clean at least once a year. Clean your knives, secateurs, spades, hoes, etc. after every time you use them.

You used to be able to use Jeyes Fluid for this, but these days use an alternative – even just soap and water is a big help.

Compost

Compost is often the reason why people do not get good crops and it is usually seedlings that suffer. Partly rotted compost is a menace. Be sure your heap is a hot one, which is made by building up a lot of material, and layering grass clippings with newspaper to stop it getting all messy.

If your compost isn't hot, sterilise it by putting it over a hot brazier for a few minutes and then allowing it to cool before using it. Otherwise you are simply spreading microbes about the garden.

HORTICULTURAL FLEECE

Before we look at controlling pests it is worthwhile looking at the use of horticultural fleece. This material is strewn over your crops, or attached to hoops that fit over raised beds or parts of the garden. It has several important properties for the grower of perfect produce.

Essentially, horticultural fleece is a very fine net, so small that insects cannot get through – even the tiniest ones. However, it still allows light, water and air to penetrate, enabling the crop to grow without molestation from flying insects. If you tuck the fleece in under the soil it is pretty good against rabbits too.

PESTS IN THE GARDEN

There are any number of pests in the garden, but we should look on them as fellow inhabitants of the world. Try not to eradicate everything or completely wipe out a single population of any pest. The whole point of managing the garden is to keep a natural environment that allows us to achieve our goals without creating a pristine, sterile space. So do try to allow a little leeway when it comes to eradicating pests. I like to leave a cabbage here and a carrot there for the wildlife to keep a toehold on life.

Polytunnels, by virtue of their increased size, can get extra humid and create microclimates that encourage the growth of fungal infections. Humidity is the biggest problem for plants because spores can germinate and fungi grow in nooks and crannies. Aim to keep the structure well ventilated and the temperature as even as possible, as this is a very good form of preventative control. This is always a good idea where edible crops are being raised as it reduces the need to use sprays.

Seedlings, like all children, have undeveloped immune systems. They can fall prey to a series of diseases called 'damping off'. You can avoid this by keeping the seedlings as cool as possible without restricting their growth, not sowing too many seeds in a tray and not overwatering. Also keep their temperature even, no sudden drops or heights, and use a little copper-based fungicide in the water. Some seeds are treated with fungicide already. Similarly, botrytis can affect all plants and the same rules apply about their care. You can help defeat this by dusting with yellow sulphur powder.

When you put plants under stress, by not having enough water for example, they can get powdery mildew. Keep plants well watered without splashing everything and increasing the humidity. When plants get water stressed air bubbles can form in the tubules inside the stem, a result of the tube splitting under the pressure caused by the leaf. This scar is then a site where the fungus can gain access to the plant.

APHIDS

Green, black, grey, woolly and a huge amount of trouble in the garden, aphids come in all shapes and sizes and cause all sorts of problems from wilting plants to fungal and viral infections. They use their mouthparts to pierce the tubes in the stem that carry sugar from the leaves to the rest of the plant. This syrup is under high pressure and gushes through the insect and pours all over the plant as a liquid known as honeydew.

Aphids can be found on any number of vegetables from lettuce to cabbage – anything with flesh soft enough for their mouthparts to pierce – and once a lone insect lands on its host it reproduces to produce an infestation within hours.

They remain over winter as eggs and appear as small adults in the spring. The first adult is wingless and she gives birth without mating to dozens of smaller versions of herself. When the numbers build up, some of these offspring produce wings and fly off to another host, often mating on the way to lay eggs.

Some aphids produce alternate generations of wingless and winged forms. By these methods they very quickly take advantage of the new vegetable growth in the garden.

Problems caused by aphids

The boring mouthparts of the insect cause damage inside the plant and in the process they break many water-bearing tubes in the stem. If enough are pierced the plant begins to fall over or wilt. Honeydew promotes fungal infections as spores fall on the plant from the air and grow rapidly in the sugar. Some botrytis and filamentous grey moulds are attributable to aphids. Grey mould becomes a particular problem on tomatoes and cucumbers grown in greenhouses and polytunnels. Also, plant viruses are spread around the garden by aphids. Economically this is significant, and can knock out many crops – particularly fruit.

Dealing with aphids

The easiest way to deal with these insects is to rub them off with your thumb and finger, but it is perhaps a little messy. Spraying them off with a high-powered jet with a hint of simple soap in the water dissolves the waxy covering on the insect and they simply dry out and die.

Lacewings and ladybirds eat aphids, and you can buy enough of these animals to treat a small infestation. You can also encourage the natural populations of these predators by placing 'insect hotels' around the garden.

CATERPILLARS

I tend to pick caterpillars off and carry them to a sacrificial plant where they can nibble away and pupate. It is amazing that a cabbage recovers once the larvae have gone, though it looks a little like a colander, and it still produces a deal of edible leaves. However, these are not really suitable for storing.

By far the best way to protect your plants is to cover them with horticultural fleece. This material allows enough light through for growth but is a perfect barrier from insect pests.

RED SPIDER MITES

The red spider mite casts an everlasting stranglehold on plants both inside and outside. They are difficult to remove once established but you can use derris, insecticides and little parasitic wasps to deter them.

ROOT FLIES

The carrot and parsnip, and many others, are susceptible to root flies. These are wonderful because they can smell their favourite host for miles, and they fly a few feet from the ground down the gradient of 'smell'. Their big problem is that they cannot fly any higher, so if you build an approximately 60 cm high wall around your plants the poor flies will go on by without bothering them.

SLUGS AND SNAILS

If you ask a gardener about their worst enemy you might well hear them cry about slugs and snails. They eat everything and there is a small army of them. No! There's a large army of them. We're told not to poison them with pellets and if we go out at night we only ever stand on them in our slippers and retreat in disgust.

The truth is that the average garden has millions of them, and they are probably the most successful animal we come across, save perhaps cockroaches. Dealing with slugs and snails is at best difficult. The problem is that there are just so many of them, and no matter what method you use to eradicate them, unfortunately you will fail.

Knowing that slugs and snails are everywhere, the only thing left to do is some sort of barrier control. There are many you can use, with various degrees of success. I have found eggshell a reasonable barrier, but it loses its effectiveness after a week or so because of the acidic nature of the rain. Thankfully I like eggs!

Similarly, copper rings do well. I stand plant pots on top of copper rings and they stay reasonably clear of slugs until the copper erodes and then you have to give it a good scrub with a wire brush to work once more.

A product that uses a similar approach is a battery-powered electric fence for plant pots.

VINE WEEVILS

Built like a rhinoceros, the weevil is almost indestructible as it forces its way through the garden, munching on almost everything it finds.

Hardly known 50 years ago, the vine weevil has come into prominence because of the number of container-grown plants that have over the years been sold from garden centres. Weevil attack is increasing not only in tubs and patio plants but in polytunnels, beds and plots. As more plants are grown abroad and shipped into the UK this problem is going to increase over the next few years, particularly as new species arrive from abroad that have no real predators.

The adult weevil feeds on leaves of almost every plant going, but hardly eats enough to cause anything more than an unsightly semicircular notch. However, the larvae feed voraciously on roots and can damage plants sufficiently for them to die. They particularly like young roots and consequently spring-growing plants are particularly susceptible.

Most species of vine weevil are completely female. All the eggs laid by these individuals are genetically identical to their mother. They are flightless but very good at walking and consequently spread around the garden easily. They can climb the vertical faces of flowerpots – even the plastic ones.

If you see an adult, or notice a leaf with a little semicircular notch on it you can guarantee there are weevil eggs in the soil near the roots. The symptoms of larval attack are yellowing leaves and wilting plants that do not respond to watering. You are almost sure to have lost the plant, but you might try shaking the compost off the roots and replanting.

The grubs are frequently dug out of the soil. They are greyish white and almost invariably hold themselves bent double. I must confess that if I kill anything in the garden I make sure it is not wasted. Simply put the grub on the path and by the time you have gone inside a bird will have eaten it.

Vine weevils can be controlled in a number of ways. You can try planting sacrificial plants that the grubs prefer to eat. Including primula, polyanthus and cyclamen, when planted around your crops they will attract the adults away from your growing area.

There are a number of products that use nematodes to kill grubs in the soil. Steinernema kraussei works at temperatures down to 8°C while Heterorhabditis megidis works in slightly warmer conditions. Products such as Nemasys are widely available and work well if you make sure you follow the instructions.

Finally, because the adults are nocturnal, leave traps of cardboard egg boxes for them to rest under during the day. It is surprising how many you can catch.

WHITE FLY

The first of the year's whitefly are seen in March and they attack all kinds of plants from tomatoes to cabbages. You can spray them using derris or some other insecticide or blast them off with a soap spray.

RESISTANCE

On the whole it is fungi that cause the major problems in the organic garden. There is little that can be done about bacterial infection and even less about viral, so buying seeds and plants that have high resistance to infection is one of the best ways of keeping your losses to a minimum.

The following crops have a high degree of disease tolerance. This list is not exhaustive, but you should be able to buy disease-resistant plants if you read the catalogues and seed packets.

- Brussels sprout: Exodus F1 – a high tolerance to ring spot and powdery mildew

- Carrot: Resistafly F1 – resistant to carrot fly, as its name suggests

- Courgette: Defender F1 – excellent resistance to cucumber mosaic virus

- Cucumber: Futura F1 – copes well in humid conditions and is resistant to powdery mildew

- Dwarf bean: Atlanta – resistant to halo blight, anthracnose and bean common mosaic virus

- Kohlrabi: Olivia F1 – resistant to most fungal infections

- Leek: Malabar – large, broad, body with good rust protection

- Lettuce: Sangria – viral resistant with a good amount of mildew resistance. I like the coloured leaves

- Mangetout pea: Delicata – good resistance to mildew and fusarium

- Parsnip: Countess F1 – resistant to canker and cavity spot

- Pea: Ambassador – tolerates damp and associated powdery mildew

- Spinach: Tirza F1 – one of the best mildew resistant varieties I have grown

- Tomato: Shirley F1 – tolerant to a wide variety of typical diseases, viral attack and even the occasional cold spell

However, even the best stock will not be perfect. Experiment by growing all sorts of plants, some old varieties, some new. Find out which you seem to have that special affinity with and grow the best. More importantly than anything else, grow a lot of varieties. There are a number of excellent reasons for this. First, if you lose one then, chances are, you will not lose the other. Secondly, we need genes! Don't simply stop using the old varieties because they are yesterday's stars. Their properties will very likely be really important to future generations of gardeners.

FUNGAL DISEASES

Fungal diseases are recognised by one of the following:

- Spots of red or black on the leaves

- A wet mess of stems and leaves that resembles slime

- Curling leaves

- Bruised or small fruit

- Tiny white, grey or blue hairy patches, often leaking liquid

- Moulds like mushrooms growing from the plant

- Rusty stains on crops like onions

Most plants are attacked by fungi, first and foremost. Pierre-Marie-Alexis Millardet developed a mixture that was not only the salvation of the French wine industry, but is wonderful in the organic garden. Bordeaux Mixture, a combination of copper sulphate and lime, protects plants from most fungal infection, particularly potato blight, tomato blight, rusts, peach leaf curl and apple canker.

Strictly speaking, Bordeaux Mixture is not considered to be truly organic, but it is so mild on the ecosystem that even the Henry Doubleday Research Organisation sells it. If you make a half-strength solution and spray any susceptible plants every three weeks during their growing season, you should remain fungus free.

Changing the local environment also helps. Keep plants aerated, with plenty of air between leaves, and if you should need to water, do it at ground level – not over the leaves. Fungal spores like moist, warm conditions, so deny these and you will likely avoid too many problems.

Finally, know your plants. If you see anything untoward, like potato leaves with large black spots, remove the plants and spray with Bordeaux Mixture. Finally, never compost material that you consider has been diseased, and for the same reason keep a good regime of cleanliness: don't walk on soil, don't use the same pair of secateurs to prune right across the plot without disinfecting them between each plant.

BIOLOGICAL PEST CONTROL

The bio-warfare method of pest control uses one organism to eradicate another without affecting beneficial species or the environment as a whole. Normally, and importantly, there is a minimum temperature below which these applications do not work.

Applications consist of wither parasitic wasps, microscopic nematodes that infect their host and kill them or some other deadly predator, such as ladybird larvae. The controls for whitefly, aphids and red spider mite are prepared in weekly batches to ensure complete freshness, and need to be applied on the day of receipt.

You can get biological controls for a large number of pests including:

- Aphid

- Chafer grub

- Leatherjacket

- Maelybug

- Red spider mite

- Sciarid fly

- Slug

- Vine weevil

- Whitefly

Most of these products work best in confined areas: in a greenhouse, single raised bed, cloche or cold frame.

CHAPTER 4

TECHNIQUES: CLAMPS, CELLARS AND SHEDS

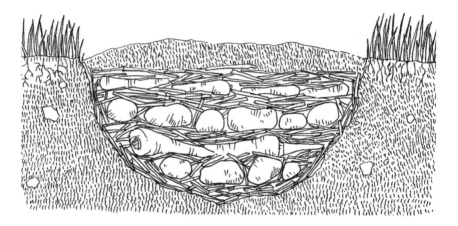

This chapter is about storage and using the plant's own immunity to keep itself fresh. These methods are good so long as we can keep the crop in optimum conditions, where it can neither dry out, be attacked by hungry animals nor be spoiled by our over comfortable homes where it is often too warm for proper storage of food.

Please note: storage is not the same as preserving.

PREPARING VEGETABLES FOR STORAGE

Having harvested your crop, as well as giving thanks and marvelling at the wonders of nature, you should be in the exciting position of deciding exactly what you are going to do with your bounty. If you are anything like me, you will have been planning all season what to do with it!

Preparing your vegetables for storage is as important a job as anything else, and should always be your first priority. Obviously, it is important to handle your vegetables and fruit with care, and it is well worth remembering what has been discussed in previous chapters. There is something in the way you hold produce that not only ensures that it isn't harmed, but that tells you something about it. Watch a celebrity chef working: he or she caresses the food, never throws it about. Handling of food indicates a relationship (something any French gardener would understand), an affinity for what has been grown.

Treating food harshly results in damage leading to rot and spoilage. Consequently, if stored in the wrong conditions, this rot can be passed on to the rest of the crop.

CLEANING

Clean your crop before storage. This can be as simple as wiping away soil or, in worst cases, actually washing. Only ever wash root crops, and leave them individually spaced to dry naturally. Washing can be avoided if your soil is not too wet, but it isn't always possible to produce clean crops, especially if you have very heavy soil. I always grow produce for show in fresh compost, avoiding the dark, cold miserable clay that makes my garden.

Cleaning your crops provides the opportunity to look for blemishes and signs of attack. You can spot slugs, insects and their eggs that will, some time approaching the spring and just when you need to rely on your stored crops, burst into life and feast themselves on your produce.

I must say, I rely on a manual inspection of my harvest rather than the addition of chemicals before harvest. No matter how often I am told hard that large organic chemicals are safe, and break down to simple, safe ones, I personally don't believe it.

CURING

Curing was mentioned in Chapter 2, but it deserves lengthier discussion. Longer-term storage is helped by allowing the produce to set itself for the following year. Almost all plants, save leafy crops, harden their outer skins to some extent, and in different ways. This is encouraged by leaving them in the air for a couple of days, in the dry. Curing happens in the soil for most crops as a mechanism against nibbling insects, and a modicum of protection from damp and fungal infection.

Curing also brings with it some changes inside the produce. For example, whatever remains of the sugar in a potato is turned to starch. Starch is insoluble in water, and if microbial attack is to work it has to be turned into sugar – and not so many organisms can perform this trick. For this reason, if your potatoes are going to ruin, they will do so within a couple of weeks after harvest.

All starchy crops benefit from curing.

CLAMPS

The clamp was used until the 1960s on farms everywhere. It was the standard way to store root crops. Huge clamps used to be dug on farms, mostly for cattle feed during the winter. In some parts of the UK it is still used to store turnips for sheep winter feed, but it has generally been abandoned.

Potatoes, carrots, swedes and mangels were all clamped, to be cut open and sealed again. They were often laid on top of the soil, with a lot of straw all around, on top and underneath and then earth shovelled on top of the piled roots, doing away with the need to dig a ditch. This saved time but that wasn't the only reason. A wet field might well fill any ditch with water and consequently ruin the crop. So it was important to know your land. Gardeners often confuse horticulture and agriculture; techniques used on farms do not necessarily work in the garden. However, the clamp is one of the exceptions.

There are many forms of clamp, or root vegetable storage ditches. The fundamental idea isn't rocket science because the best place for roots, naturally enough, is under the ground. John Seymour, the author of many self-sufficiency books, always said this was the reason why the Irish grew potatoes instead of wheat. You could leave potatoes in the ground and no bailiff would come and dig them up. But wheat needed harvesting, and once stored can be more easily taken away.

Clamps are good for storing potatoes, swedes, turnips and carrots in large numbers. Almost any root crop can be clamped, but many don't need this treatment. For example, turnips, with the help of a cloche, can be grown at almost any time of the year, and in the hungry gap you will always have enough to spare. I have also grown turnips by sowing them in containers indoors in October with a crop by March in the greenhouse. A small crop, yes, but big enough and you can store what you have grown on a shelf.

Why not store root crops in the ground?

The big problem with simply leaving your crops in the ground until you need them is that this is really only a short-term solution. Particularly in the case of potatoes, but also with most roots, a number of things occur under the ground.

Firstly, as soon as the rains come our subterranean harvest falls prey to slugs and snails. They eat most of the roots in the garden. You will be startled to find out that the average garden has millions of slugs and snails living in the soil. So your first problem is likely to be slug damage.

Furthermore, as the temperature falls, plants protect themselves by toughening themselves, some add antifreeze, others shrink and still more turn to mush at the first frost.

The clamp deals with these two problems by using a lot of straw, firstly to deter molluscs and other spoiling creatures, and secondly to add insulation so the roots are kept cool, but somewhat protected from the worst of the frosts. Straw also has the benefits of not rotting easily and helping the stored crop to remain dry.

A clamp on a smallholding

If you think about it, there is not a lot of difference between a clamp and a drainage ditch. Consider where you are going to site the clamp. If you are lucky enough to have a field drain, then you can happily site your clamp downhill of the drain. This is because the drain itself diverts water from above, leaving the ground below much drier.

Making a large clamp

Dig a trench: 1 m wide, 30 cm deep and as long as is needed, which is usually one-tenth of the length of the rows of crops that produced the roots in the first place. Fill your trench with straw and pile the dry, cleaned roots in layers on top. Separate each layer with handfuls of straw, until the whole mass is about 80 cm high.

Cover the outermost layer of straw with a layer of soil at least 20 cm thick to seal the clamp. If you live in a very wet area, you can cover your clamp with plastic sheeting when it is raining, but leave it open to the air for the most part.

Using a clamp

All you have to do is scrape away the soil, then the straw and retrieve the food you need, then carefully reconstruct the straw and soil when you have finished. You should, as far as possible, collect the roots on a dry day.

Making a box clamp

Box clamps are traditionally filled with peat, though any number of alternatives can be used. It is a large box with a layer of dry peat on the bottom, and then alternating layers of, say, carrots with layers of peat. Finally the whole lot is topped with peat.

Instead of using precious peat you can use sterilised soil, which you can prepare by placing a spadeful of soil on a tray over a fire – I use a brazier for this purpose. You can also use spent, sterilised compost or dry sand. You should splash out on silver sand for storing vegetables.

One of the fundamentals is that a clamp is dry, but retains enough humidity to keep the produce fresh and to stop the roots from over curing themselves. Obviously, you never really get as good as same day fresh, but if properly stored, deterioration is minimal.

These days it is common for people to use plastic boxes, but for a good clamp you really need wood. We used to use old-fashioned fruit boxes for storage, but you can't get them these days. Plastic doesn't allow the contents to evaporate moisture sufficiently, and spoilage from fungal attack can result if they get too wet. A slatted wooden box of around 50 cm is ideal for an individual crop; you can fit a couple of month's worth of carrots in such a container.

Make sure you put the box in a dry, dark, cool place.

CELLARS, LARDERS AND SHEDS

Root cellars

The root cellar is an American idea where an element of cold is required. The root cellar is an almost 2 m deep hole in the ground, lined with wood, to all intents a shed buried in the ground, usually with steps going down to the door. However, you can simply use a cellar, so long as it is a dry one. Root cellars are only really useful where the drainage, or lack of rainfall, makes the void dry, but cold. Almost anywhere in the UK, a 2 m hole will fill with several inches of water in no time.

Root cellars vary in construction. You can find them built into the side of a hill, while others are half dug in the ground and half shed built on top. Sometimes they are dug in a barn, with access via a trapdoor. Importantly, no brickwork is used unless it is supported by a damp-proof course, and most are made from wood.

Usually root crops are spread on shelving or mini-clamped in boxes of earth or sand. They were also used to store salted food, from vegetables to fish.

Home cellars

Many of us are blessed with a cellar, but often these are not brilliant places for storing because of the damp. The other problem is you find many slugs and insects in cellars, and the most important inhabitant is a great fat frog, who will happily grow even fatter by eating the other occupants.

Cellars with earth floors are better than those with stone floors because they allow the house to breathe, and the atmosphere in the cellar is less damp. This is the best kind of cellar for storing.

Wooden shelving and wooden boxes for storing are best, and if you can, try to keep each apple, potato and what not lying singly on shelves with a good deal of air circulating. Storing in bulk should be done, if possible, in dry sacks.

If you have a freezer in the cellar, keep it away from any stored crops on shelves because condensation from the heat exchanger can increase humidity in the immediate vicinity, and the increase in temperature can cause crops to go off quicker than you would have thought.

Larders

It seems strange having to explain larders these days. They are a cold room with plenty of ventilation, usually on the coldest corner of the house, next to the kitchen.

When I was a boy we had a larder in our council house which was next to useless. The problem with most larders of the type was a lack of ventilation and they were frequently too warm. You should be able to feel the coolness of the larder when you walk into it. In order to maintain the temperature, you should be sure of an excellent, draught-free door. Good larders have a substantial stone floor and plenty of air.

If you are lucky enough to have one you should set the space with dairy, meat, fish, fruit and vegetables, wine and beer, each in their own sections. There are two problems in poorly organised larders. First, the possibility of contamination is most obvious. Food on shelves can drop, drip or contaminate food below. The second is insects, which can get in and spoil the store food. This is most easily dealt with using horticultural mesh.

A larder is no place for cooking, or storing warm food. Always complete these operations in the kitchen before taking the various products into the larder for storage. Again, as with the cellar, keep the freezer away from the rest of the other foodstuffs, preferably at the coldest part of the room.

Sheds

Sheds can be a nightmare for storage. They can be a hiding place for rats, almost invariably damp and insecure when it comes to human interference.

If all you have is a shed, make sure it is as watertight as you can manage it, and wrap your stored produce in newspaper to keep out the damp and vermin. Lockable steel boxes, filing cabinets and old fridges make for good storage containers – make sure that the humidity does not build up by opening the doors from time to time.

When I had a shed as my only storage solution, I used to keep roots in sacks, hung from the main roof apex. This makes them fairly impervious to rats and mice, but I had to make sure the roof didn't leak down the string that was holding the sacks in place. An extra layer of roofing felt was enough to ensure this.

Another problem with sheds is temperature. They get very hot during the day, very cold at night. This constantly fluctuating temperature is bad for storage and is the major reason why sheds should be avoided if possible.

CHAPTER 5

TECHNIQUES AND RECIPES FOR PRESERVING FOOD

In this chapter we look at the various techniques for cooking and preserving food. In it you will learn the basic techniques for bottling, salting, making jams and preserves, steaming, juicing and syrup making.

This chapter focuses on preserving rather than simple storage, and here all the culture of humankind is to be found. People think culture is about art, sculpture, paintings, music, and of course, it is. But there is as much history and culture, invention and art in a jar of pickle. People have invented these things over thousands of years, modified and improved them. The fact that we have such a wonderful array of fine preserves is because of the inventive art of peoples long forgotten. When you preserve food, you make history and are connected right back to the beginning of agriculture.

USE YOUR SENSES

Any of the techniques for preserving food can go wrong. From time to time you get a spoiled bottle or can of food on the supermarket shelves, and they have been produced in a so-called safe, foolproof environment. So don't trust your preserved food. Look at it for signs of spoilage such as black bits or other discolouration. Then smell it – are there any unpleasant odours? Listen – did gas escape when you opened the container? Have a small taste – a tenth of a teaspoon. Does it taste good? If not, discard it.

Always cook your stored food thoroughly, making sure it is thoroughly heated through to at least 75°C for several minutes. Preserved food needs proper cooking, nothing rare or undercooked will do.

SALTING

Salting is an ancient method of preserving food, which relies on an excess of salt to draw water from the food and so making life impossible in its tissues.

Plants for salting should be in perfect condition, washed, dried and any leaves or coverings removed. You need a large pot with a lid – earthenware is the most frequently used. In most cases the food is layered; a layer of produce and then a layer of salt, until the vessel is filled. Then put the lid in place. Broad beans are one such crop stored in this way, though I have also seen cauliflower and aubergines.

Aubergines used to be bitter and needed to be soaked in salt water to draw out bitter juices. Preserving them in salt served two purposes, preserving and tempering. However, modern aubergines are not so bitter!

The large amount of salt in the preserving of these vegetables draws water from them, which can dissolve some of the salt. It is important, therefore, that the vessel is checked for air gaps, and that more salt is used as necessary.

When the produce has been in the salt for a number of days, it is more or less in its final salted state, wrinkled and tiny due to water loss. When you are ready to use the product, simply take it out of the vessel and soak it in water. It will take 24 hours and a number of changes of water to get the saltiness out of the food.

I have to say that runner beans stored in this way and cooked after soaking are superbly tasty.

PICKLING

This is a step up on the evolutionary scale from salting. When making pickle, the osmotic material is usually sugar, and then vinegar is used to hold the preserved state. However, in the case of onions and shallots, salt is used to draw water from the bulb, and then this liquid is replaced with vinegar.

Pickled Shallots (or Onions)

The art to this is peeling your shallots under water to stop the chemicals turning your eyes to a reddened, agonising mess.

Makes about 1 kg

9 kg shallots

300 g salt per 1 litre water

2 tablespoons mustard seed

1 tablespoon allspice

2 tablespoons black peppercorns, crushed

2 tablespoons cloves (whole)

1 teaspoon ground ginger

2 bay leaves

5 cm piece of cinnamon stick

1 litre vinegar

- Peel and wash your shallots. (This step usually makes you cry.)

- Prepare a salt brine solution: add the salt to the water and warm to about 30°C. Leave the shallots to stand in this solution for 24 hours.

- Make a pickling vinegar by adding all of the remaining ingredients to the vinegar and leave it to stand overnight.

- Remove the shallots from the brine, dry and then add to the vinegar.

- Pack into sterilised jars. Seal and leave for at least 6 months. Stores for 1 year.

Pickled Gherkins

The process for pickling gherkins is similar to that for pickling shallots, but there are some variations. Some people sprinkle the gherkins with dry salt instead of soaking them in water. Some people add a tablespoon of sugar to the spice mix because gherkins can be awfully bitter when pickled.

CHUTNEYS

Chutneys are sugar and vinegar based, but usually spices are added.

Green Tomato Chutney

This recipe is good for green and red tomatoes, and usually at the end of the season you have a combination of red, yellows and greens. You can leave out the raisins, but the basic idea is useful for making all types of chutney.

Makes about 3 kg

2 kg green tomatoes, roughly chopped

700 g onions, finely chopped

450 g Bramley or other cooking apples, peeled, cored and chopped

200 g raisins

500 g golden caster sugar

600 ml white vinegar

25 g fresh root ginger, finely chopped

1 level teaspoon mixed spice

2 teaspoons salt

- Place the tomatoes in a pan with the other ingredients.

- Heat gently, stirring until all the sugar has dissolved.

- Raise the heat and bring to the boil. Then simmer for 1 ½ hours until thick and smooth.

- Pot in sterilised jars and seal immediately and label when cool. Allow to mature for 4–5 weeks for the best flavour to develop. Stores for 6 months to a year.

Sweet Brown Pickle

This is the pickle to make right at harvest time, setting aside the fruit you need. The process is to cook the food in a preserving vinegar, and the osmotic element comes from sugar, which also serves to provide sweetness. The vegetables can be chopped as finely as you wish and they can be prepared in a food processor for a very finely textured pickle.

This pickle can be made with the cut offs from damaged fruit and vegetables, but be sure you only use good stuff in the actual mix.

Makes 4 × 450 g jars

250 g carrots

1 medium swede

4 cloves garlic, chopped

125 g dates, finely chopped

2 onions, finely chopped

2 medium apples, unpeeled and diced (you can use dessert or cooking apples, obviously dessert give a sweeter flavour)

15 small gherkins or 1 medium-sized cucumber, cut into small cubes

250 g brown sugar

1 teaspoon salt

2 teaspoons mustard seeds

2 teaspoons allspice

4 tablespoons lemon juice

500 ml malt vinegar

- Put all the ingredients into a large pan and stir well. Bring to the boil, stirring constantly.

- Reduce the heat to a gentle simmer and maintain this for 2 hours, stirring every 20 minutes or so. The pickle should look moist but not wet or runny.

- Ladle the pickle into sterilised jars and seal immediately. Stores for 1 year.

Tomato Relish

When is a relish not a pickle? No answer really, except a relish does have some strong flavours, more sugar and probably less vinegar than a pickle.

Makes 2 × 500 g pots

150 ml white wine vinegar

180 g caster sugar

1 teaspoon mustard seeds

1 teaspoon salt

1 large onion, finely chopped

1 small red pepper, chopped

1 stick celery, chopped

750 g tomatoes, chopped

½ tsp chilli flakes or 1 fresh chilli, chopped

- Heat the vinegar, sugar, mustard seeds and salt in a pan, gently stirring until the sugar is dissolved.

- Add the onion, pepper and celery and bring to the boil. Reduce the heat to a simmer and cook for 10 minutes.

- Add the tomatoes and chilli and simmer for another 20–25 minutes or until the mixture thickens.

- Stir and pot into sterilised jars immediately or leave to cool completely before eating. Stores for 6 months.

MAKING JAM

Jams are lovely. The staple of any nation's poor, jam and homemade bread has kept many a family happy when there was little else to eat.

Jam relies on two things. The first is the amount of sugar in the jam to provide an osmotic preservation. That means the sugar draws water from the food, making it impossible for microbes to live in that environment. The second thing is the amount of pectin in the food determines its setting properties.

If you are at all worried about your jam setting, use preserving sugar, which has good grains for melting and contains pectin. Once you have brought your jam to the appropriate temperature, using preserving sugar, you should get a set.

Usually a temperature of 106°C or 200°F is about right for setting jam. You can use a jam thermometer, but generally if the mixture is boiling robustly it will set. Clearly these temperatures are high enough to kill any germs in the jam.

Testing for set jam

You can test whether your jam has set by drizzling a teaspoon onto a cold saucer. Allow the jam to cool a little and then, using your finger, push it. If it crinkles, you know your jam is ready.

Sterilising your jars

Your jars should be sterile and hot when you ladle your jam into them. A wide-necked jam funnel makes this task easier, though you need to sterilise this too.

You need to sterilise the lids as well as the jars. You can place the lids in boiling water, for at least 15 minutes. I usually place the jars in the oven at 175°C/350°F/Gas Mark 3 for 15 minutes, having thoroughly washed and rinsed them.

Sealing the jar's neck

Pour the molten jam into the jar right up to the neck and leave it for a few seconds to heat the airspace above. You can add a wax disc, laying it gently on the surface of the hot jam. This melts the wax, causing a seal and further protecting the jam. Seal the lid tightly and allow the jam to cool. As it cools, so will the air in the neck of the jar, causing a partial vacuum. This holds the lid even more strongly in position and pulls the rubber seal in the lid fully in contact with the rim.

The science of jam

The process of making jam creates a syrup that will osmotically draw water from the food and, indeed, any microbes that fall on the jam. If you are able to get hot sterile jam into hot sterile jars and put sterile lids on top, you will have a product which should stay perfect inside until it is opened. However, lids do not seal properly every time and so it is possible that in a batch of ten jars of jam, one of them might go off.

Most jams are 1:1 sugar to fruit, though some jams need more sugar. Normally, 1:1 sugar is not enough to keep some microbes at bay – particularly penicillin. Once you have opened your jam it is liable to go off in a few weeks. Fortunately jam doesn't last that long in our house!

As a caveat to this, honey is 80% sugar and lasts for thousands of years, but nectar – the substance with which bees make honey – is only 1:1 sugar to water and ferments rather rapidly.

Pectin is a protein that forms a colloidal gel in the presence of sugar and will set in its own time, as many jam makers know all too well. If your jam doesn't get a setting point, assuming it is hot enough, the usual answer is to add more sugar.

Safety first

Jam burns. When the sugar crystallises, the energy that is kept in the liquid is given off as heat, usually on your hand. Wear protecting heatproof gloves and have some cold water ready for you to thrust your hand into just in case.

If you get a burn bigger than a 50p, seek medical attention; these burns might run deep and can ulcerate.

Upside-down Jam

This is our quick fire method of making jam, which is simple enough for making small quantities of jam. It works with any fruit, just make sure there are no stalks or hard bits in the fruit and wash your fruit before use. Try to use perfect fruit only when making this jam.

1 kg fruit makes about 4 jars. 2 kg fruit makes about 8 jars
Equal quantities of fruit to sugar. Use jam sugar for best results setting-wise

For up to 1 kg fruit: juice of ½ lemon and an equal amount of water

For up to 2 kg fruit: juice of 1 lemon and the same of water (use the lemon halves to measure the water)

- Put the sugar, lemon juice and water in a pan over a gentle heat and allow the sugar to begin to dissolve.

- Add the fruit and move around in the dissolving sugar. Don't stir vigorously.

- When the sugar starts to colour with the fruit juice, turn the heat up to medium and stir gently, ensuring all the sugar has dissolved.

- Bring the mixture to the boil and boil for about 5–10 minutes. Test for setting by lifting the spoon high above the pan and watching for the droplets to slow down. When the drops are setting the jam is ready.

- Leave to cool in the pan for about 10 minutes. Then stir and ladle into sterilised jars. Leave to cool and label and date the jars. They should store for at least 3 months in the cupboard. Once opened, keep refrigerated.

MAKING MARMALADE

I know exactly what you are thinking: I don't grow oranges. That's the purist coming out in you. Thankfully, the bliss of Seville oranges is that they cannot be anything but seasonal! The point is that they arrive in January and you have to make marmalade with them. Many people preserve food bought at the supermarket, particularly when it comes to making jams and preserves and some pickles. And why not?

How to freeze Seville oranges

Rather than making a batch of marmalade that has to last the year through, why not freeze the oranges and make smaller batches as you need them. Prepare the oranges as freshly as possible as they will lose some of the pectin during freezing, but the fresher they are the more pectin will be retained.

Wash the skins with lightly soaped warm water and dry well. Place them in freezer bags in batches of 1 kg so it is easy to defrost the amount you need later on. Freeze immediately and quickly.

You can also freeze them after cooking and before adding the sugar. Simply put them in a lidded container and freeze until needed.

Things to remember when making marmalade

- Make your marmalade as soon after purchasing the oranges as you can, to ensure the freshness and therefore the best flavour.

- Because it is difficult to give an accurate weight for citrus fruits in a recipe, an average weight is given below – a 50–60 g difference won't matter. With large grapefruit, you may need to use half a grapefruit to reach the required weight.

- Wash the fruit with lightly soaped warm water before use, even if it looks clean.

- Prepare the fruit over a dish or bowl to catch all the drips of precious juice.

- If you need to use a muslin bag to hold bits and pieces of fruit, when it comes to lifting it out, use tongs and squeeze out as much of the liquid as possible. It will look slimy and gel-like; this is the pectin that will help set the marmalade.

- When the marmalade is cooking it often produces a scum on the surface. This can be dispersed by adding about half a teaspoon-sized knob of butter and stirring. The scum will disappear.

Seville Orange Marmalade

Makes 8–9 × 450 g jars

1.5 kg Seville oranges

2.5 litres water

2.2 kg sugar

Juice of 2 lemons

- Wash the fruit. Cut each orange in half and squeeze out the juice. Put the juice in the pan, without wasting any. Put all the pips in a dish ready to be tied in a muslin bag or square.

- Scoop out all of the pith and flesh from each orange and put it in the pan. If there are some very thick bits of pith (most will dissolve in cooking and help the marmalade set), remove them and add them to the pips.

- Cut the peel into strips as thin as you like to have in your finished marmalade. Add to the pan with the other orangey bits. Then add the water.

- Put the pips and pith bits in a muslin bag and tie well. Pop them in the pan also.

- Bring to the boil, then turn down the heat and simmer for 2 hours until the peel is very soft.

- Remove the bag of pips with tongs and squeeze out all the juice. You will see that it is quite slimy – this is the pectin being extracted.

- Remove from the heat and stir in the lemon juice and the sugar. Return to a low heat and stir until all the sugar has dissolved.

- Bring to the boil and boil for 10 minutes. Then test for the setting point.

- When ready, leave to cool for 10 minutes and then stir to distribute the peel evenly. Pot into sterilised jars and label. Stores for 6 months.

MAKING FRUIT JELLY

1. Preparing the fruit

   Wash the fruit well and discard any with mould or that are very badly damaged. There is no need to discard bits of stalk as this will be done during straining.

2. Cooking the fruit

   Place the fruit in the preserving pan and just cover with water. Some recipes may call for a little more water, such as blackcurrants and quinces which take more water. Stir in the lemon juice at this point if it is included in the recipe. Bring slowly to the boil, then simmer gently until the fruit is tender. Each kind of fruit takes a different amount of time to cook, but it must be tender or the juice won't be easily extracted during straining.

3. Straining the cooked fruit

   Arrange the jelly bag or muslin securely on the stand or whatever you are using and place a large bowl underneath. Ladle the cooked fruit into the bag and leave to strain. This will take between 2 and 10 hours depending on the amount you are making and the type of fruit used. When it is ready, the pulp should look dry and the juice will have stopped dripping through.

4. Measuring the juice

   The juice must be measured, as it is the only indicator of how much sugar to use. Use a large ladle to transfer the juice into the jug rather than pouring it in from the bowl.

5. How much sugar?

   For every 100 ml of juice, add 90 g of sugar. This is the basic and easiest way to measure the sugar and gives better setting results. For a less sweet and less firm set, use 75–80 g sugar per 100 ml of juice. Use white granulated sugar as this gives the clearest finish and best flavour to the jelly. To help the sugar dissolve at its quickest rate, place it in an ovenproof dish or shallow roasting pan and heat for 10 minutes at 140°C/275°C/Gas Mark 1.

   Warm your sterile jars at the same time. Turn off the heat when you have taken out the sugar; the jars can be removed just before you need to fill them. If you stand them on a baking sheet it is easier to lift them in and out of the oven.

6. Adding the sugar

Pour the juice back in the pan and bring slowly to a fast simmer. Then turn down the heat and add the sugar. Stir well over a very low heat until all the sugar has dissolved. Check the back of the spoon for sugar crystals to make sure.

7. Boiling and setting

Bring the mixture to the boil and boil gently rather than too vigorously – it will still set. Check for a setting point after 8 minutes as you would with jam, using a cold plate or saucer and dropping onto it a small amount. Allow to cool then push with your finger: if it wrinkles and remains it is ready. Boil for 2 minutes more if it is not ready. While some fruits are boiling, a scum develops on the water's surface. Skim the surface under the scum with a large flattish spoon or fish slice and lift it away.

8. Potting the jelly

Use a preserving funnel to help with the potting. Remove the warm jars from the oven and immediately ladle the jelly carefully into the jars. Do this slowly as if it is done quickly bubbles can appear in your jelly, spoiling the look and the air will eventually spoil the jelly too. Seal with the lids immediately to ensure that the jelly stays fresh for the longest period possible.

9. Labelling

This is a very important step in all preserving, but especially with jellies as the only clue you have to the contents is the colour. So label them clearly with the type of jelly and the date made, which enables you to keep a check on how long it has been kept.

BOTTLING

If you walk through the supermarket you will find row after row of wonderful bottled food that were once the realm of the homemaker, but have become a standard must buy. Modern cooking is populated by examples of stored ingredients becoming main meals. Yet foodstuffs and ingredients, like baked beans, can be made at home and are far tastier than any factory-made product.

I get much more excited about a container of asparagus soup hiding in the freezer than some sloppy old asparagus that I then have to do something with to make edible. This, to my mind, is the best way to use our crops: to make food, real dishes. What better way to store apples than in apple pies in the freezer?

Bottling preserves food through a combination of killing germs by boiling, storing them in such a way as to keep germs out and finally, making doubly sure your food isn't spoiled, by adding a preservative, usually salt for vegetables, sugar for fruit.

Preparing for bottling

You will need the following:

- Jars or bottles of whatever type you have or prefer
- Racks for draining
- Large pan for cooking and later boiling the containers
- Oven gloves
- Sterilising solution if you prefer

Use perfect produce

Only bottle perfect produce. You can cut away the bad bits if it is obviously safe to do so, but don't try to rescue bad produce – there is no point.

How to sterilise bottles

There are many ways of doing this. First, wash the bottles. They need to be completely clean and free from taint – and you would be surprised at how long it takes to get the smell of vinegar out of a pickle bottle.

You can use Milton sterilising solution in liquid or tablet form, which basically bleaches the germs away. It is not so kind on the lids, so it is probably best to boil those. Those who worry about this method of sterilising for environmental reasons should consider that it takes a lot of energy and causes a lot of pollution to heat jars to boiling. Unless you have your own wood burner, which is carbon neutral for the most part, there is really no other method so environmentally friendly and effective.

Boiling bottles

This is possibly the method that gives you the most confidence – this way, you really know that the bottles have been completely sterilised. Put about 2.5 cm (1 inch) of boiling water in the bottom of the jars and place them in the oven at 160°C/325°F/Gas Mark 3 for 20 minutes. This will completely kill any microbes.

The seal on the lid is more important than anything else. If you are using Kilner jars, check the seal, which is usually removable, and replace it if it is warped or damaged. If you are reusing jam jar lids, pay particular attention to the lid seal; it is usually only very small. I don't reuse jam jar-type lids more than twice. You can buy replacements very easily.

Sterilising Kilner jars

Kilner jars or similar look the business when you are storing anything, but they have a number of issues that I find a bit awkward. First of all, you have to be sure you remove the rubber seal and sterilise this separately. After dozens of attempts to make them completely sterile, I have resorted to soaking them in strong brine. Obviously, boiling water would be a good method, but the rubber is mis-shaped by the heat. I tried using sterilising tablets, but they went a strange colour, as though they had spent the year in the desert, and I worried about the seal and my ability to keep them clean.

How to prepare food for bottling

Anything you are going to keep in a bottle has to be boiling at the point of putting the food into the bottle. You have to have your glassware at boiling point too, so it does not crack when the hot food goes in. As everything is hot, be careful: wear the appropriate protective clothing and use oven gloves to protect your hands.

Also, have a bowl of cold water available, just in case of an accident when you might need to plunge a scalded hand into water. This is especially important when making jam. Don't forget to have dry hands before touching anything likely to be hot – water transmits heat so very easily.

Fill the jars to the top of the neck, leaving the screw top lip (where the lid will go) empty. This is sometimes referred to as the head space. Screw or fix the lids in place and then pop the jars into boiling water and boil for 40 minutes. The jar lids should be amply covered with boiling water for all of this time. If the water needs to be topped up, always do so with boiling water from the kettle.

After this period, carefully drain the water and leave to cool. Most bottled materials prepared in this way will last for a minimum of 8 months.

Bottled Fruit

Makes 2–3 jars

500 g sugar

1 litre water

2 kg fruit

- Prepare a syrup by dissolving the sugar in the water. Bring to the boil, stirring all the time.

- Remove stalks and other debris from the fruit and wash.

- Sterilise your bottles and lids – preferably by boiling them. Then add the fruit to the jars and pour over the boiling syrup. Close the lids. It is really important that the lids seal properly, which is why Kilner jars are good as they give you a click method for ensuring that this happens.

- Place the bottles in boiling water to their necks. Boil for 20 minutes, which is plenty long enough to sterilise the contents. Stores for 6 months.

PASTEURISING

Why pasteurise when you can just boil?

Pasteurisation heats food to 75°C for 20 minutes, or even longer. Boiling heats food until it boils, often well over 100°C, and this process has several implications. First, proteins denature at temperatures above 40°C and at boiling have so degenerated many of them resemble their original state. Thus, vitamins can become useless by overheating. There is a fine line between reducing the food value and removing the poisonous bacteria. Pasteurising strikes a balance between the needs of taste and nutrition on the one hand and safety and health on the other.

The other problem with boiling is that flavour changes, and whereas this should be nothing compared to having a safe product, it does have its drawbacks. You can get rapid chemical reactions that make your cordial bitter, for example, and frequently there is a film or debris left behind after the harsh boiling process.

It has to be said that boiling in many instances is important. Food keeps longer, and you have no worries about the safety of your food.

Louis Pasteur, the French scientist who invented pasteurisation, carried out experiments to find out what temperature and time combinations were necessary to kill 99.9% of the bacteria and fungi in food. He found that 100°C killed all the bugs, and that increasingly lower temperatures needed longer times to kill the germs.

To be sure, I always pasteurise at 75°C and for 30 minutes, which seems to work for apple juice for instance. However, this is not sterilised: there will be some bugs inside the apple juice or product and you should not keep it for longer than a couple of months. That said, if I freeze apple juice I always pasteurise it first.

How to pasteurise

Pasteurising machines come in many forms. Some are thermostatically controlled to control the temperature exactly. Others are little more than double boilers with a thermometer. You can use your own pans to pasteurise, placing the containers in water at the appropriate temperature, but this is not accurate so, if you can afford one, it is much easier to use a machine designed for the job.

A thermostatic pasteuriser takes all the effort out of the job of maintaining the correct temperatures for successful pasteurising. Having used many Heath Robinson affairs to pasteurise juice, I have found the convenience of a dedicated machine to be the best.

Simply fill the bottles or preserving jars with the juice or food to within 2.5 cm of the lid. Put the sterilised lids in place and stack the jars into the rack in the machine. Fill the machine with water 2 cm from the lips of the jars and set both temperature and timer. You may need to wait for the required temperature to be reached in calculating the pasteurising time – some machines compensate for this time while others do not.

Remember that 75°C is very hot, and can burn very badly – especially when lifting jars and/or racks out of the machine. Make sure you use excellent protective gloves.

CORDIALS AND EXTRACTING JUICE

It is surprising how many people think that cordial and drinks in general are only made in factories as a pleasant drink, and ignore the fact that cordials are a perfectly good way of preserving food. They are in fact an ancient way of preserving fruits (and some vegetables and herbs).

Essential vitamins are destroyed by boiling and the ancient way of making cordial was to boil the fruit, add the sugar and then store it in stoneware containers with lids. Modern cordials, however, do not have to be boiled; pasteurisation and the use of sugar produces a healthier product with greater vitamin concentration. Making apple juice, or any fruit juice, is an excellent way of preserving the fundamental goodness of the fruit, and leaving you with some pretty useful pulp too.

Hard fruits, like apple, should be diced but not peeled and then put into a pulper. This can be something as simple as a mortar and pestle, and I frequently put the fruit in a muslin bag and beat it to death. You can, and this is the easiest way, put the apple pieces into a food processor and pulse it. Add the juice of 1 lemon to every kilo of apples.

From here, put your pulp into a muslin or a jam bag or some other fine strain and press to separate the juice from the pulp. Leave the juice overnight to settle. It is best if you can put them in sterilised, sealed bottles on a stone floor – floorboards wobble as you walk around the house and keep the juice from clearing.

Remember that everything you use in this process needs to have been sterilised.

Using a steam juicer

This is by far the best way of juicing anything from apples to strawberries. It relies on steam to extract the juice, and moreover you can add sugar to your pulp in situ – this will then make for a really good preserve. It is a remarkably simple system: in essence you pass steam through fruit, the steam condenses and the juice falls into a reservoir, to be collected.

There are a number of products available. I use the Hackman 3 in 1 because you can use it for other things such as making jam or as a general steamer for vegetables. You need to be very careful about the temperature and wear protective gloves at the very least, as everything gets very hot. I would also suggest you put newspaper down to collect spillage, which is fairly inevitable, especially when dismantling the steamer.

One of the really exciting things about this way of juicing is you can add the appropriate amount of sugar when the juicer is in action. You layer fruit and sugar in the steamer part of the extractor. This is then steamed and the juice is collected in the reservoir. The initial juice is laden with sugar, so this is poured back into the steaming basket to even out the sugar concentration.

HOW MUCH SUGAR

- If you are freezing the juice, it doesn't need any juice.

- For juice kept in the fridge, use 200 g per kilo of fruit.

- For juice kept at room temperature, use 500 g per kilo of juice.

STARTING OFF

Wash your fruit and discard any that are bad. Place the fruit in the steamer. Load the bottom pan with water, about 4 litres is enough, and assemble the extractor: base pan, juice container, steamer and lid. Bring to the boil on high heat and then turn the heat to a strong simmer (sometimes known as a rolling boil).

It takes up to 30 minutes to get the juice to flow – you can check in the tube to see if there is any juice. Then the juice is simply decanted into sterilised bottles, which are filled as full as possible and sealed with the lids. Pour the first litre or so back into the juicer in order to even out the sugar content, if you have used sugar at all!

Bottled juice will store for 6 months.

STEAMING APPLES AND OTHER FRUITS

It is possible to steam apples to give up their juice. I have found it best to chop the apples into small pieces, around half a centimetre in size, which can be a little time consuming. You will find that 4 kilos of apples will produce about 2 litres of great apple juice.

IMPORTANT NOTE

Steam juice extractors get really hot, and so is the juice inside before you pour it off. Most steam juice extractors have a plastic tube to draw off the liquid as it forms inside the reservoir. This also gets really hot! So as with everything in preserving, make sure you wear protective gloves and have some cold water on hand.

Flower water

This type of machine is perfect for making perfumed waters. Stack the petals of your favourite flowers in the extractor/steamer, and after a short time you have some wonderful waters. Lavender, rose, apple, calendula and honeysuckle all make wonderful scented waters, the bases for many perfumes and health drinks as well as cosmetics.

You have to clean the machine well after use or else everything smells of roses – not a bad thing I suppose!

Making juice the old fashioned way

This is simple really. You are using a food processor or a crusher to pulp down the cell walls of the fruit, thus releasing its contents. This is then put into a press to remove the liquid juice. It needn't be expensive. You can buy a plastic bucket with an electric drill-driven pulper for around just £20, and presses can be found for not much more.

Essentially you bash the life out of the fruit and press the juice out. What could be easier? This method is mostly used with hard fruits like apples and pears, and semi hard fruits like plums and damsons, although you do need to de-stone them first. Equally, stoned fruits do well with steam extraction.

What to do with the pulp

Eventually there will be lots of pulp left behind. Don't throw it away! If you make sure there aren't any stalks or bad pieces of fruit, you can then use the pulp for other things such as these fantastic granola bars. There are plenty of other recipes for granola and, yes, you can freeze the pulp to continue to make this brilliant bar, time and time again.

Granola Bars

Makes approx. 5

100 g fruit pulp (apples, strawberries and raspberries are my favourite)

300 g porridge oats

50 g butter

2 tablespoons sugar

- Preheat the oven to 180°C/350°F/Gas Mark 4.

- Mix together all ingredients.

- Bake for 30 mins and then cool.

FRUIT PURÉE AND BUTTERS

Purée

Treat all fruits, apples to strawberries, more or less the same way: heat the fruit pulp with sugar until it falls, that is, when the juice starts to run out of the fruit and it loses its structure. The amount of sugar varies according to the fruit – the sharper the fruit, the more sugar you add. Test the sugar content by taste, though as a rough guide use about a third of the whole weight of the fruit in sugar.

Fruit that browns when cut, like apples, will need lemon juice added to it. Fruit that doesn't have any juice will need a little water.

Apple Purée

Makes 1 small jar

250 g or 2 large cooking apples, peeled, cored and diced

Juice of ¼ lemon

80 g caster sugar

4 tbsp water

- Simply cook all of the ingredients together at the same time on a low heat. Stir frequently to avoid the sugar or fruit burning.

- Once the apple is soft and mushy, push through a sieve to get the desired consistency.

- Place in sterilised jars. Stores for 1 month or for 1 week after opening. If pasteurised it will store for 3 months or it can be frozen for a year.

Fruit butters

Fruit butters differ from jam in that they are not set with pectin, but rather are a loose preserve that is not completely set. Neither are they stewed for as long as a purée. They are cooked in large quantities, 5 kg at a time, which is ideal at harvest when you have such a lot of fruit about. Again, we will use apple as an example, but the basic idea is the same for whatever fruit.

Apple Butter

This makes a great addition to ice cream and it can also be used as a base in apple pies.

Makes approx. 2 × 1 kg jars

5 kg apple

500 ml water

Juice of 2 lemons

Cinnamon (optional)

- Peel, core and dice the apple and add it to a heavy ovenproof pan. Add the lemon juice and water – about 500 ml to 5 kg apple.

- Place the pan in the oven at around 100°C/210°F/½ Gas Mark and cook for 6 hours.

- After about an hour, and then on the hour, give it a stir. If you like spices with your butter – in the case of apple, cinnamon is good – then add it after 3 hours.

- At the end of this time, bottle as you would jam. Stores for 6 months or 1 week after opening.

Note: No sugar has been added, and you might like to add it halfway through, but it is not necessary and is not a part of the recipe.

VACUUM PACKING

Used for a number of years in supermarkets, vacuum packing is now becoming widespread in the home. It can be used to stop food spoiling by excluding air and producing an environment in which food can be kept safely.

It is also common to cook food in the bag using hot water. If you are able to get the contents really hot, you can cook the food, making it bug free and able to be kept for weeks or even months depending on the contents and the storage conditions.

Buying a vacuum-sealing machine

There are a number on the market, and in their most basic form they consist of a pump and a heating element. They are all fairly robust and can be used on a table top or in a wall mounted position. Before you buy, make sure you are able to get a ready supply of the bags you need for that particular machine.

The cost of the machine is not the most important consideration. You are not, for example, going to save a lot of money in a short time with this machine.

Using the machine

In the simplest way, you bag the food, place the open end of the bag into the machine and press the handle down. The machine pumps out the air and then the heating element seals the bag. The food is now in an airtight environment, with the contents sealed.

If you are going to freeze the bags of food, blanch it first as described below. But you can also cook the food, in individual portions if you like and store these separately, but be careful the freezing process does not turn it to mush.

FREEZING PRODUCE

There are so many ways to freeze food from harvest, and in some ways this is the ultimate storage method for much produce. But it comes at a price, not in financial terms, but environmental. The freezer is on all the time, forever, burning away at the electricity, producing carbon dioxide and degenerating the planet.

To be honest, it was the banks of freezers in the supermarkets that inspired this book. What would happen I thought, looking at a pile of sawdust and rotting food from a leaking freezer, if we had no power – we'd have to store our food the old fashioned way. That said, no modern book on storing produce can be complete without a look at freezing.

Organising your freezer

The problem with freezing certain produce is that it works its way to the bottom of the freezer and remains there to be discovered only when you defrost the freezer. This is a real problem because, well, what's the point? You have spent money freezing something you don't eat and that has taken up precious space that could have been used elsewhere. Try always to freeze food in containers, so you can put similar items together and so that nothing escapes into some distant corner.

Keep your freezer regularly defrosted. Don't allow ice to build up – especially since ice just creates fossils of food in some frozen permafrost deep inside. Ice makes the space smaller, the cooling less efficient and the freezer more expensive.

Don't overload your freezer. Make sure there is plenty of space and that the contents do not stick together during freezing. It is more economical to run two freezers with no overcrowding than one packed really tight.

Steps for successful freezing

BLANCHING

There are a number of requirements for successful freezing. First of all, you need to remove as much of the surface bacteria and bugs as possible. This is done by blanching, which reduces the spoilage load, that is, the number of microbes available to spoil your food.

In order to blanch, simply 'dip' the food in already boiling water. You are not cooking it and you will need to monitor the blanching time carefully. As a general rule, the softer or thinner the food, the shorter the blanching time necessary. So something very hard like a piece of carrot might need 90 seconds, whereas something soft and thin like a thinly-sliced piece of courgette might need only 45 seconds.

Blanching is made easier if you place the food in a container such as colander, and insert this carefully into the boiling water. When the time is up, remove and plunge into iced water (cold water will do, but it is a second best) to reduce the temperature and stop any cooking. Remove the food as soon as it is no longer warm. Then dry and pack into a freezer bag, clearly labelling it to indicate what the product is and when it was frozen.

FREEZE IMMEDIATELY

Then the food needs to be frozen as quickly as possible. Set your freezer at its coldest setting for a couple of hours before use. The quicker the food is frozen, the smaller the ice crystals that form. When water crystalises, the size of the water crystals in the cells of the food can often rupture the cells, causing damage, often in the form of mushiness. The smaller the crystal the less the damage – so always freeze as quickly as possible.

What a freezer cannot do

In a strict sense, freezing does not preserve food, it simply slows its rotting. Bacteria caught in ice do not die as such; they simply stop reproducing. Having defrosted, some spoiling bacteria will reproduce at a quicker rate than they would normally, and consequently frozen food should be cooked and eaten as quickly as possible.

It is important to maintain the temperature of your freezer – don't be forever opening the door. If it defrosts by accident, you really ought to consider if your food inside has been compromised and whether it needs to be thrown away.

Can I freeze fruit?

Fruit with a high water content does not freeze well, although many other types of fruit do. Gooseberries, currants, cranberries, blueberries, rhubarb, apple pieces and grapes all freeze well – but usually they taste different when thawed. Defrosted fruit is not the same as fresh fruit, and consequently it is often a good idea to freeze fruit products rather than the fruits themselves.

Strawberries, for example, are notoriously difficult to freeze. The crystals make them all mushy and the result is very disappointing. However, you can freeze strawberry purée (see Chapter 7 for the recipe) which can be used to flavour cakes, yoghurt, cakes and ice cream, or just eaten as it is.

Don't blanch fruit

Because it cooks so quickly, I tend not to blanch fruit unless I am making a purée or fruit butter. In this case, I put the fruit in a colander and simply dip the lot in boiling water for a few seconds.

Sprinkle with sugar

Many fruits do well, that is, keep their flavour better, if sprinkled with sugar before freezing. This goes particularly for blackcurrants, to which I add 1 tablespoon per kilo, sprinkled evenly over the berries. I then pack them into a freezer bag, removing most of the air before freezing. They are wonderful spooned into ice cream or just consumed having been defrosted.

Sugar syrup

Sugar syrup can be added to some fruits before packing for freezing. Boil 500 ml of water and carefully add 500 g of sugar. Allow to cool and clear. Place the fruit in sterilised jars and pour over the cool syrup to the neck, making sure there are no air spaces. Seal and boil in a water bath for 15 minutes.

I particularly love to cover cooked rhubarb with sugar syrup before freezing. Once defrosted you can then pour the syrup over the dessert. This also works very well with raspberries, and if you allow the raspberries to stand in the syrup for a couple of hours before freezing, you get some of the flavour in the syrup.

DRYING FOOD

It is possible these days to take your kitchen a thousand miles south where there is the perfect climate for drying food. Drying vegetables and fruits makes great snacks, preserves food and is about the only time in the UK where you can use oil as a way of keeping food free of outside infection. I am thinking, of course, of wonderful home-dried tomatoes – could there be any better food?

Drying or dehydrating food is the oldest method we have for preserving food, and might well date back over a million years. We might not know the origins of drying food but it is realistic to believe it is very ancient indeed.

Not in the oven

Yes you can dry food in a slow oven, as though you were making meringue, only slower. The problem with this method is the material gets too hot, curls up, loses much more than just water and, generally speaking, produces good results only sporadically.

The best way, apart from in the sun, to produce dried food is to use a purpose-built machine. This way you will have consistently good results and guarantee that your food loses only water, not vitamins, or becomes too tough. Most machines have a heating coil and a fan, which pushes warm but not hot air around a series of shelves, dehydrating the food on the way.

There is no limit to what you can dehydrate – fruit, vegetables, herbs, mushrooms – and you can either make food to eat in a dehydrated form or to re-hydrate later.

Fruit snacks

Fruit snacks are a wonderful way to have a healthy sweet. Dehydrate say, an apple, cored and sliced, so that it is chewy but not brittle. Test the fruit for chewiness every half an hour after the first 3 hours of drying. Since it only loses water, the flavours intensify, the vitamins are still there and the snack is rich in goodness.

Muesli

Try drying your fruits and adding them to oats to make a wonderful cereal. A combination of apples, pears, grapes and apricots makes a wonderful start to the day – even better when it all comes out of your own garden!

Drying vegetables

Vegetables such as carrots, turnips, swedes, anything you care to mention really, should be blanched before drying. They should then be stored in dry jars and will retain their properties so long as they are not wet by accident. They can be used directly in soups – a large bag of dry onion rings is good for this, as are mushrooms.

It is best to slice vegetables so they are no thicker than 5 mm, but if you don't mind persevering with whole vegetables, they work too so long as they fit in the dehydrator.

Always re-hydrate vegetables by soaking them in an equal volume of water. Always cook re-hydrated vegetables straight away – they have no keeping qualities.

How long to dehydrate?

In a way this depends on what you want to do with the product. When food has lost all of its water it becomes brittle. This is good for long-term storage, but not too good for some products. Most machines come with three power settings and a guide to drying times. Machines available on the market vary in power but most drying at 40°C will take a minimum of 4 hours. Once dry, products will keep in a jar with an airtight lid.

Drying herbs

Herbs are dried for medicinal and culinary reasons and, in order for their volatile oils to remain active, drying should be cool and slow. Ten hours at 20°C is common but, again, refer to your specific machine's instructions. It wasn't until I started to dry my own herbs that I realised why you often need less dried herb in a recipe than fresh.

Tomatoes

The cost of a dryer is worth the opportunity to dry tomatoes. I slice them in two and dry until they become wrinkled and a little bit pliable, even slightly moist. I then grind salt until it is very fine and spread over the dried, cut surfaces. Salt lightly at the rate of 10 g per kilo of tomatoes.

In a jam jar add a sprig of basil and pack the dried tomatoes loosely inside. Fill up with extra virgin olive oil. Leave for a month – you will never taste better.

CHAPTER 6

A–Z OF GROWING, STORING AND PRESERVING VEGETABLES

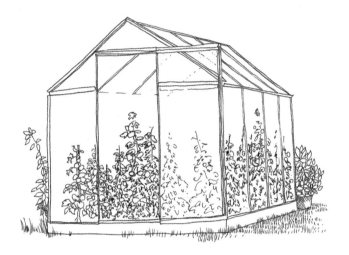

So many people do not realise that the plants we take at harvest time are not necessarily the best ones to store. This chapter shares some tricks for growing perfect crops for eating straight away, while also growing perfect harvesting crops at the same time.

GET THE GERMINATION RIGHT

What happens to a young plant can have consequences later on. One of the biggest problems is low-level fungal infections that find their way into the adult plant undetected, and hence when the plant is being stored the fungus takes over. For all intents the plant looks healthy, and is perfectly good to eat, but it doesn't last so well as others, and you are left scratching your head as to why some plants do better than others.

Problems in the later life of crop plants start in the early growing days when seedlings are easily infected. This is usually down to too much water and too little ventilation, and the problem called 'damping off' raises its ugly head.

This is basically a fungal infection and it occurs in propagators where the combination of humidity and temperature is just right. Such plants appear to be uninfected, but they could easily have low level infection causing problems later on. I find this problem particularly in onions, where plants coming from infected seed trays in propagators turn out to be poor storers though they didn't show signs of problems at harvest.

The easiest way to avoid this problem is to be sure there is enough ventilation in the trays and that the temperature is not too high. This is called 'growing hard', making the seedlings work for a living, and consequently their immune systems are improved. Keeping them hot and what might be said to be over watered will grow plants quickly, but not with such brilliant immunity.

This chapter includes notes for growing for storage and sometimes special notes about harvesting for storage too.

ASPARAGUS

A perennial vegetable, asparagus plants die back right down to soil level and re-sprout the following year. It is necessary to plan the growing site for asparagus carefully as they are long-term crops and will continue to crop for 15–20 years.

Growing for storage

Asparagus is a difficult plant to store, but you are best storing those plants that were produced a week after the first spears. Remember you are only ever taking about 60–70% of the plants anyway, so a couple of spears per plant is about tops. Store those spears collected on the same day.

Asparagus needs to be kept moist, so ensure beds are regularly mulched and soil is loose. Add fertiliser in early spring and again after harvesting, but do not overfeed. Support growth with canes and remove any seedlings as soon as they appear. In autumn, cut the plant down to 2.5 cm (1in) from the soil and remove the dead foliage.

Harvesting

To ensure the longevity of the crowns, asparagus spears are not usually harvested until the third year. Cut spears singly, using an asparagus knife, below the soil when they are 14–18 cm (5–7in) tall.

How to store

NATURALLY

Store in dry silver sand in a deep box, in a dry but cool place. They should store for up to 2 months.

FREEZING

Blanch the spears in boiling water for 2 minutes before cooling in iced water. Pat dry and place in a freezer bag and freeze as quickly as possible. They should keep for 12 months. Good in soups, but not so good whole.

ASPARAGUS IN ASPIC

Aspic is the jelly that comes from pigs' feet, but you can use gelatin instead. Bring your asparagus to the boil and put it in a sterile dish (a long, asparagus-shaped dish is good). Pour over this a jelly made from 1 litre of gel with 3 level teaspoons of salt. Make sure the gel is boiling hot and pour over the stored asparagus. Tap so the bubbles are removed before the jelly sets. Keeps for about 2 weeks in the fridge.

ASPARAGUS CONFIT

This is similar to aspic, but uses duck fat to cover the asparagus. Melt the fat in the oven, get it very hot and pour over the prepared asparagus. Stores for up to a year in the freezer, but the tissues go mushy making it not bad for soups.

Pickled Asparagus
Good for about 30 asparagus spears
30 asparagus spears

300 g salt per litre of water

Pickling vinegar (see the Pickled Shallots recipe p.49)

- Peel and wash your asparagus and trim the ends. Prepare a salt brine solution by adding the salt to the water and warming to about 30°C. Leave the asparagus to stand in this solution for 24 hours. Every now and then give the basin a tap to remove air bubbles.

- Make a pickling vinegar and leave it to stand overnight.

- Place the asparagus in sterile jars – don't wash the salt off them. Cover with the prepared vinegar.

- What you get from this product is a slightly tangy, almost asparagus-flavoured (you do get a hint of asparagus), slightly mushy stick. Stores for a year.

AUBERGINE

Grown in greenhouses and polytunnels, aubergines have the reputation of being terribly bitter, but modern varieties are much less so, producing only the sweetest fruits. Originally, aubergines were soaked in a salt solution overnight to remove the bitterness but these days they do not need this treatment. Storing in salt used to kill two birds with one stone: combating the bitterness and preserving the fruit.

Growing for storage
Don't let your aubergine plant lean on the side of the pot – cushion it with a little straw. Avoid fungal infections by keeping the humidity as low as possible. I tend to use ring culture pots.

It is always best to ensure these plants never do without water and nutrients. In this respect, treat them a little like tomatoes. Water them at three-day intervals and feed weekly, using a high potash feed. They prefer light, well-drained soils – I usually grow mine in compost. Certainly avoid cold, heavy clay soils.

These plants need to be kept warm, particularly for germination and the early growing stages, where they need 16°C as a general minimum.

Thin flowers to one per stem to ensure large fruits.

Harvesting

Fruits can be taken when they reach their full colour. Store only those plants that are perfect, and I have found those taken early by a day or so are better in storage. Aubergines very quickly become overripe, so letting them ripen off the plant increases their storage time by a couple of weeks in the fridge.

How to store

NATURALLY
Aubergine is not an easy plant to store naturally. Keep at less than 10°C but certainly no colder than 4°C. They deteriorate very quickly in cold conditions, and this is not a food you can freeze easily.

FREEZING
Cut into cubes or slices and blanch in boiling water for 2 minutes. Plunge into iced water and then pat dry. Place in freezer bags and freeze quickly. They should last 6–8 months. I have found the longer you keep them, the mushier they become.

SALTING
There are a number of salting regimes you can use for aubergines. Slice or dice the fruit and place them in a colander. Evenly sprinkle all the surfaces with salt at the rate of two teaspoons per kilo. Leave in the fridge on a plate so any drips can be collected. These will last a fortnight.

PICKLING

There are lots of recipes out there for aubergine pickle but they do not include much in the way of sugar or salt as a preservative. Consequently there is not much in their keeping qualities.

However, ratatouille is perfectly good for storage in Kilner jars on the shelf or in the freezer.

Ratatouille

Double up this recipe if you need to – it all depends on the number of jars and amount of storage space you have.

Makes 2–3 × 500 g jars

2 courgettes

1 medium red onion

1 red pepper, seeds removed

1 yellow pepper, seeds removed

1 aubergine

5 ripe tomatoes

2 tablespoons olive oil

3 cloves garlic

Salt to taste

Method 1

- Chop all ingredients into small pieces.

- Cook on the hob until everything is boiling and soft.

- Bottle in the usual way in sterile jars and boil for 20 minutes. Stores for 6 months.

Method 2

- The more traditional way is to roast the ratatouille in the oven. Set the oven to 180°C/350°F/Gas 4.

- Chop the vegetables and put them in a roasting tin. Coat everything with olive oil – I use my hands to do this.

- Roast for 20 minutes.

- Ladle into sterile jars and boil for 10 minutes with the lids sealed. Stores for 6 months.

BROAD BEANS

Beans of all kinds are easy to grow, and easier to store! There come in a range of varieties from long-pods to the dwarf cultivars, so there is a broad bean to suit any plot garden.

Growing for storage

I prefer to start my beans in pots in October and overwinter them in a cold greenhouse. I then plant them out in April, and they grow very quickly from then on. They mature slightly earlier, thus avoiding black-fly.

I also sow outdoors in May, covering this crop with horticultural fleece to avoid insect pests. Generally speaking, this is the crop I store.

Broad beans prefer a well drained but moist soil, rich in nutrients. Although the members of this group (legumes) can use nitrogen from the air in soil to make their own proteins, they generally do much better in soil that is quite fertile.

Plant them in 'double rows'. That is, plant each about 40 cm apart, then plant another row about 30 cm further along from the first. Another double row can be planted about a metre away. This way the plants support each other and grow more evenly.

Harvesting

When pods ripen, collect them regularly to prevent tough beans. You will see the beans bulging inside the pod and, as soon as the pod starts to change colour, they will be ready.

You can allow them to dry in the pod if the summer is dry. They will be ready for storage (see the section Natural below) when they do not dent on biting.

How to store

NATURALLY

Slowly dry your beans in a container – a large Kilner jar will do the trick. Place about 2 inches of dry rice in the bottom and then fill with beans. The rice will absorb any moisture left in the beans.

I usually boil the beans before shelling because it is easier that way.

FREEZING

You need to shell the beans and blanch them for about 2 minutes in ½ kilo batches. Then plunge in iced water, dry, bag and freeze as quickly as possible. However, beans keep so very well when dry, there is little need to freeze them.

Bottled Broad Beans

These make a great standby – if you don't mind them mushing a little.

Makes 2–3 × 500 g jars

  1 kg broad beans

  6–10 cloves garlic

  3 level teaspoons salt

  1 teaspoon sugar

- Boil about a kilo of beans in water for 10 minutes and then remove the skins.

- Return to the pan on a low heat and add the garlic, salt and sugar, stirring well for 11–15 minutes.

- Ladle into sterile jars with some of the hot liquid, packing as many beans (and the garlic cloves) into the bottles as possible. Place in a pan of boiling water, having secured the lids, and sterilise for 10 minutes.

- Keep in a cool dark place. They should last a year, but I have never kept them longer than 6 months.

HARICOT BEANS

This covers French beans, white beans, butter beans, flageolets and kidney beans – all of them easy to grow, all of them wonderful to eat.

Growing for storage

This group will continue to crop, on their support, from late June to September. You should store those that appear in August.

All they need is weeding, feeding and watering, and you should mulch them with good compost once the plants are 30 cm high. Water very regularly and keep the weeds away – they find it very difficult to compete.

Harvesting

Snip off the bean pods when they are of a decent length. If the skin becomes hard and rough, it is too late. It is best taking them early rather than late. If you are going to remove the seeds, do so after a day's drying.

How to store

NATURALLY

In the fridge they keep for a month, but after that they seem to deteriorate quickly. They succumb to fungal attack, which is reduced by keeping them in an open aspect – but then this uses up a lot of room.

They last for about 6 weeks spread in a box and covered in dry sand.

SALTING

If you need to, you can salt them by simply layering up 2–3 cm of beans with 1 cm of salt in an earthenware pot. Press the beans down well as you layer them. Watch out that after a few days the liquor doesn't wash the salt from the top layers. They will last forever, but need soaking in several changes of water before use.

Runner beans are best salted, chopped into 3 cm-sized pieces.

DRYING

This is the easiest way to dry seeds for use all winter through. You simply grow your beans, of whatever variety. Leave the beans in the pods until the pod starts to shrivel. Then remove the pods and take them indoors. Keep them on a dry shelf until they look as though they will shrivel no more. Then remove the seeds and store them in a sealable jar on some dry rice. As long as they are perfectly dry, they will basically store forever. They can be used directly in food as the cooking water gives them plenty of hydration.

Dry runner beans, I think, are tough, and so I soak them in a little hot water first and squeeze off the skins before use.

FREEZING

To my mind there isn't a lot of point, but you can chop, blanche for a minute, plunge in iced water, dry and freeze in bags. They will store for a year but they lose their texture and sometimes even go mushy.

Baked Beans

By far the best way to save beans, this baked beans recipe is really nothing like the baked beans you get from the supermarkets. You can freeze this dish and defrost it to make a meal at any time, or you can bottle into jars, sterilise and keep them on the shelf.

Really a meal in themselves, these baked beans are also very tasty. They have a richness of flavour not found in even the most expensive tins of beans. In essence, the trick with this recipe is that the tastier your tomato sauce, the nicer the dish.

You don't have to use haricot beans even though the majority of baked beans are made from them. You can buy them by the bag which will last for ages, but equally you can grow and dry them yourself. You can also add dried broad beans. If you do use another type of bean, don't overcook them, because you will end up with a mush.

Makes approx. 2 kg

1 kg dried haricot beans, soaked overnight in water

25 tomatoes, finely chopped

4 leeks or red onions, finely chopped

1 tablespoon olive oil

400 ml passata

8 teaspoons sugar

Salt and pepper to taste

- Set your oven to 160°C/325°F/Gas Mark 3.

- Put the beans (which having been soaked will be plump and soft) into an ovenproof dish along with the tomatoes, leek or onion and olive oil.

- Cover with passata and add the sugar, salt and pepper.

- Stir and bake for 30 minutes. Stir again and bake for another 30 minutes or until tender.

- These baked beans can be frozen in dishes ready for heating up or bottled in Kilner jars. Either way, they will store for a year.

BEETROOT

Growing for storage

It is possible to have this easy-to-grow crop all the year round if you grow them in a combination of cold frame, greenhouse and open beds. Interestingly, you always get the same yield of beetroot per metre of crop. If you plant them close together you end up with small roots, further away they grow bigger, but if you weigh the amount of produce per metre, it is always about the same.

They are also good for growing in pots and containers.

Make sure that your roots have plenty of nutrients in the soil and that any you are going to store are not damaged.

A constant supply of beets can be had outdoors from June to late September by sowing indoors from February, then every two weeks, and sowing outdoors in late April. Continue this regime until mid July. You can sow indoors again in September. Use 'Boltardy' varieties as this little plant loves to run to seed.

A warm, sunny spot with slightly acidic soil is ideal. The soil should be finely hoed and nutritious – your beetroot has a lot of colour and flavour to make. Keep them well watered but not too wet.

Harvesting

Use a trowel to lever the plants out when they are about 3–5 cm across. Refer to your seed type for guidelines, but certainly do not let them grow bigger than 5 cm.

Cut off the leaves straight away: they evaporate water from the root. The leaves are quite edible and are especially nice in salads. Leave the roots open to the air for a couple of days to harden the outer skin.

How to store

NATURALLY
They will last for the best part of a year in dry sand, perhaps three at a pinch.

FREEZING
Peel and cut your beets into cubes about 1 cm square and blanch for about 3 minutes. Or cut them into slices a couple of millimetres thick and blanch for a minute. Then plunge them into iced water, dry and place in freezer bags. These will last all year.

Bottled Beetroot
There are lots of recipes out there that are variations on a theme. For this one, choose fresh, smaller beets, about 3 cm in diameter.

Makes 2 jars
12 small, fresh beetroots

500 ml vinegar

500 ml water

500 g sugar

2 teaspoons plain salt

- Boil the beets for 10 minutes and then peel.

- Bring the remaining ingredients to the boil and add the beetroot. Boil for another 15 minutes.

- Pour into hot sterile jars and seal, boiling for a further 10 minutes to ensure sterility. Stores for 3–6 months.

Pickled Beetroot

Makes 2–3 jars
12–15 small beets, each approx. 3 cm in diameter

330 g salt per litre of water

Pickling vinegar (see the Pickled Shallots recipe p.49)

- Boil the beets for 10 minutes and then peel them.

- Prepare a salt brine solution: add the salt to the water and warm to about 30°C. Leave the beets to soak in this solution overnight.

- Make a pickling vinegar and leave it to stand overnight.

- The next day, remove the beets from the brine, dry them off and add to the vinegar. Pack into sterilised jars and store for 6 months. They will take a couple of months to be ready. Don't worry about any discolouration – beetroot juice is an acid indicator.

BROCCOLI

The following information covers the calabrese varieties as well as purple sprouting.

Growing for storage

Sow indoors in winter and plant into their final growing positions in April, making them a long crop in the ground.

Like all brassicas, they are susceptible to club root. I counter this by growing individual plants in pots. I then add a lot of lime to the soil. I use a bulb planter to cut a plug of soil from the ground and coat this with a trowel-full of lime. I then fill the hole with fresh compost and plant in this. More than anything, prevent club root by not walking on your soil!

Make sure they are well watered and that the soil is reasonably fertile. Stop the plants from rocking in the wind – a good heeling in when the plants are mature enough works wonders. As the flowers are forming, feed the plants with tomato fertiliser. This will provide the extra boost they need for producing more florets. The smaller florets appear when the main flower head is harvested and these smaller ones are, in fact, the best for freezing.

Harvesting

Don't wait too long for the flower head to open, and take the florets regularly as they appear. They are flowers, and once pollinated they go off rapidly. Cut the stem as close to the florets as you can without cutting into the head and take them to the kitchen ASAP.

The tiny holes in broccoli make excellent hiding places for insects and grubs, so soak them for 15 minutes in slightly warm water to remove most and then prepare for use or storage.

How to store

NATURALLY

Wrap them in newspaper and bury them in dry sand. They will last for a month in this way. Taken from the latter weeks, they make excellent soup.

FREEZING

Cut the florets as close as possible, so you have pieces with a width of only a couple of centimetres. Blanch in boiling salty water for about a minute and then plunge in iced water. Dry and bag, and freeze quickly. They should be good for a year.

Broccoli Pickle

This is really fun. I like it with a teaspoon of cloves added to the pickle mixture.

Makes 2 large jars

Pickling vinegar (see the Pickled Shallots recipe p.49)

1 kg broccoli, cut into small florets

- Make a pickling vinegar and leave it to stand overnight.

- The next day, blanch the broccoli in salted water for 1 minute and then plunge into iced water.

- Then add the broccoli directly into the pickling vinegar, having knocked off the excess water from the florets en route. Pack into sterilised jars.

- After a couple of days, give the jars a good knock on the table to knock off any excess air bubbles. Repeat this a couple of times more over the next fortnight.

- Stores for a year.

BRUSSELS SPROUTS

This plant is the Christmas standby – almost everyone either loves or hates them. What applies to all the other brassicas applies to sprouts, particularly when it comes to club root.

Growing for storage

The most important tip about growing sprouts and keeping them for a while is that if they are securely heeled in the ground, the sprouts themselves will be really tight, not what gardeners call 'blown'. Then the crop has its own built-in mechanism for short-term storage – it will last at least a month in this position, especially if protected by fleece.

Sow indoors in January and plant out in their final growing positions in May. You will have large plants that will need care: plenty of water and in July a dressing of organic fertiliser. In September, heel in to avoid wind-blown rocking.

Harvesting

Take from below when the sprouts reach a couple of centimetres long and use straight away. Store only those sprouts which do not have yellow leaves.

How to store

NATURALLY

I'm sorry, I cannot bring myself to concoct recipes for sprouts! There are just some plants that have to be in season and sprouts is one of them.

They can be stored by leaving the outer leaves intact and placing them on a board, covered with a dry sack or sheet. This can then be covered with dry sand and stored in a cool dry space. They have been known to last six weeks using this method.

CABBAGE

It is possible to have fresh cabbages in the ground all the year round. A sowing and planting regime that goes from January with January King through to March with Pixie and another sowing of All Year Round in September to be covered with cloches in the bad weather, will see you cabbage-bound all the year through.

Growing for storage

There are some great foods to be had from preserved cabbage, but they also keep for about a month on the shelf. The best way to be sure to have cabbages perfect for storing is to grow them under horticultural fleece and, of course, treat them for club root in the way described above for broccoli.

If you sow in drinks cups of compost and grow on in these containers until the plant is at least 15 cm tall, you will then be able to plant out without problems at almost any time of the year. I plant out into the greenhouse in September, leaving the plants to grow cold through the autumn and winter.

Be sure the ground is free draining and fairly nutritious. A top dressing with organic fertiliser when planting out is a good idea.

Keep the watering regular and even. If you let them go without and then add some water as a remedial gesture, they are more likely to split.

Harvesting

Harvest when the heads are firm to the touch. Cut at the base of the plant, leaving a little of the stalk intact. Keep the outer leaves as a protective layer. If you leave them in the ground too long they will split.

How to store

NATURALLY

Leave them to dry off, if necessary, and then wrap with the outer leaves and place them on a shelf, with plenty of air around them. They will keep for a month in this fashion. If you have room for a permanent cropping of cabbages, place new additions at the back of the shelves and roll them all forward in succession.

FREEZING

Frozen cabbage is a great convenience and it will store for at least a year. Slice the cabbage into strips about half a centimetre thick and 5–8 cm long. Soak these in water for about 15 minutes with 1 tablespoon of salt. Then place them in boiling water for 15 seconds. Pat dry and gather into a big freezer bag for freezing. Cool quickly.

Sauerkraut

This is simple and easy to make and it is really life giving. It is really fermented cabbage with salt to control the amount of fermentation and microbes growing in the cabbage.

Cabbage

Salt

- Cut the cabbage into finger-sized pieces.
- Sprinkle 1 tablespoon of salt per 1 kg of cabbage onto the cabbage and then transfer the lot to a crock, which is usually an earthenware, wide-mouthed vessel.

- Really pack the cabbage and salt down hard and then cover with a plate. Cover the whole thing with a towel.

- After 24 hours look inside. The salt will have drawn the water from the cabbage, and the resulting brine will protect the cabbage as it ferments. Every couple of days, make sure the cabbage is under the water. If there isn't enough water, top it up with brine: mix 330 g salt per litre of water (but you won't need a litre).

- The plate needs to be washed each time it is removed, so you don't infect the kraut. If you see a mould on the surface of the water, just skim it off – it's nothing to do with the sauerkraut. After a couple of weeks it will be tangy, and after a month it will be ready! Stored in a cool place, it will last 6 months.

Pickled Cabbage

Red or white cabbage may be used in this recipe.

Makes about 1 kg

1 large cabbage

Coarse salt

Pickling vinegar (see the Pickled Shallots recipe p.49)

- Choose a firm, fresh cabbage and remove all the loose outer leaves and any that are damaged.

- Shred the cabbage as finely as you prefer. The tough inner core can be used to make soup or discarded.

- In a large bowl, put a layer of cabbage and cover with a layer of salt. Continue layering, finishing with a layer of salt and leave for 24 hours.

- Drain away the salty liquid and rinse in cold water.

- Pack into sterilised jars and cover with the pickling vinegar. Seal well and label and date the jars. Stores for 6 months.

CARROT

Like potatoes, carrots come in early, main crop and late varieties. I tend to stick with what I know best, and choose Early Nantes for early carrots: sow in March – August and harvest July – October. Then I sow Autumn King in March and keep it going until November. Then I sow Artemis F1, a newer carrot that will overwinter quite happily.

So it is possible to have carrots in the ground all year round and there are truly hundreds of varieties to try.

Growing for storage

Good free draining, not too fertile and very finely dug. It is no exaggeration: you can hoe the soil all day long and that is exactly how they like it. Cover with horticultural mesh to keep the carrot fly out and you have the perfect micro-climate for growing.

Easy to care for, carrots more or less grow themselves. Watch for carrot fly damage. If you keep them at bay using horticultural mesh, all you have to do is thin the carrots as they grow.

It is a good idea to water in drought times because if they dry out they will swell when their next water arrives, and the carrots will split.

Harvesting

They are best harvested when they are of an appropriate size, that is, when you want them! But I think that they are at their best when around 2–3 cm in diameter. Always lift the carrot with a trowel or fork; never pull by the leaves.

Once lifted, leave to cure in the open for a couple of days, having removed the leaves to cut down evaporation.

How to store

NATURALLY
Carrots store well in boxes of dry sand (silver or river sand is best). You can layer sand and carrots until your box is full – wooden boxes are the best. This way they will last easily for 6 months in dry, cool conditions.

FREEZING
I freeze baby carrots and some of the narrower varieties such as Mokum F1. Pick them early and they are delicious. A bag full is ideal in the freezer and will store for about a year. Blanch for a minute, plunge in iced water and then freeze in the bag – leave a little bit of leaf and they look wonderful.

DRYING
Yes, you can dry carrots. Cook them first, as this makes them keep their colour. Simply peel, and grate the carrots and pop them in a steamer until cooked. Remove excess water and then I use a desiccator to dry them completely.

The carrots can be used in making carrot cake, carrot bread, soups and stews and they also store perfectly in a jar for about a year.

JUICING
You can use a steamer or a juice extractor and then freeze or store the juice for up to a year (a fortnight once opened). I prefer to use a pasteuriser to ensure the juice is safe. If you don't pasteurise, boil your carrot juice – the flavour will change but the contents will be sterile. Expect settling of solids in the juice.

Pickled Carrots
You can use carrots in the Brown Pickle recipe (see Chapter 5), which is gorgeous and stores for ages, or you can make a carrot pickle, which is also tasty.

Makes 3 × 500 g jars

Per 1 kg carrots:

400 ml pickling vinegar (see the Pickled Shallots recipe p.49 or simply add 1 teaspoon of bought pickling spices to 400 ml vinegar)

200 g sugar (I prefer dark brown, but any will do except for jam sugar)

- Cut the carrots into small cubes and steam them until cooked. Have the hot, sturdy, sterile jars ready.

- Bring the pickling vinegar and sugar to the boil. Add the cooked carrots and simmer for 30 minutes.

- Add to the hot sterile jars, secure the lids and boil for 15 minutes in a water bath. Stores for a year.

CAULIFLOWER

Some people insist that cauliflowers are really difficult to grow and that they never have much success with them. If you think of them as being thirsty plants that don't like water – that is, make sure they always have water available to the roots, but the soil itself is free draining – then you will do better.

Growing for storage

Apart from watering, like all brassicas they need to be heeled in so they don't rock about in the wind. Space them wide apart, at least 45 cm, and weed carefully between the plants on a regular basis. Protect with horticultural fleece, as cauliflower can be susceptible to the various infections nibbling insects will bring.

They do best in a fairly rich soil. Add plenty of good quality compost to the soil, and improve the drainage. An extra top dressing of organic fertiliser about 6 weeks after transplanting gives them a growth spurt. I treat them like all brassicas, growing in drinks cups, and with plenty of lime.

Harvesting

Take them as soon as possible after the curd (flower) is formed, when it is a perfectly light cream/white. Once pollinated, the heads darken and look less appetising very quickly.

How to store

NATURALLY

Cauliflower won't last more than a couple of weeks on the shelf. If you cut them into small pieces and vacuum pack they will keep for around a month, especially if kept in the fridge.

FREEZING

Wash the florets in salt water to remove any insects and then towel dry. Then blanch in boiling water for a maximum of 1 minute. Plunge into iced water and then towel dry before bagging and freezing quickly. Stores for about a year.

DRYING

It takes some time to achieve this, but cauliflower florets do dry well in a desiccator. You can then put them in a jar on top of some dry rice and later use in soup – this is a really good storage method and they will last forever as long as they are dry.

PICKLED CAULIFLOWER

There are dozens of recipes for pickled cauliflower with lots of variations. Of course, piccalilli has lots of cauliflower in it. You can also add cauliflower to the Brown Pickle recipe given in Chapter 5.

A simple cauliflower pickle, with no other ingredients than cauliflower and vinegar, is hardly made these days, the vegetable's delicate flavour is lost in the vinegar. In the pickles mentioned above, the cauliflower portion is included for texture.

Cauliflower Pickle

This light pickle will only last for a couple of months and should be kept in the fridge.

Makes about 3 jars

330 g salt per 1 litre water

1 kg cauliflower florets

450 ml vinegar

- Make a brine solution by adding 330 g of salt per 1 litre of water.

- Wash your florets. Blanch in boiling water for 1 minute and then plunge into iced water.

- Boil 50 ml of water and add the vinegar. Transfer the florets from the brine to this mixture, tapping away the excess brine.

- Pack into sterilised jars. Then seal and boil for 10 minutes. Stores for 6 months.

Piccalilli

This is a traditional pickle that I could include under any vegetable in this section really, but it probably deserves to be listed under Cauliflower more than anywhere else. It is a simple pickle, though to make perfect piccalilli you need to get the balance between the vegetables just right. This comes with experience really, so the key word is experiment!

Makes approx. 6 large Kilner jars

3 kg mixed vegetables of your choice, all finely chopped (you will probably include 750 g silverskin onion and 750 g cauliflower. In addition, try gherkins, cucumber or courgette, carrot, turnip, celeriac, celery, pak choi stems – the list can be endless)

Salt

1 level teaspoon turmeric

4 teaspoons dry mustard powder

3 teaspoons ground ginger

780 ml white vinegar

Extra white vinegar

500 ml vinegar of your choice

1 tablespoon cornflour

- Lay the prepared and chopped vegetables on a tray and, instead of soaking in brine, sprinkle with salt. Leave overnight to sweat.

- Mix together the spices. Then mix them with a little white vinegar to make a paste and keep to one side.

- In a large pan add all the vegetables, the spices and 500 ml of the white vinegar. Bring to the boil and add the remaining 280 ml white vinegar and 500 ml of another vinegar of your choice. Simmer for 25 minutes.

- Make a paste with the cornflour and a little white vinegar.

- Stir into the pickle mixture and boil for about 2 minutes.

- Ladle the contents into sterile jars and tighten the lids. Stores for a year.

CELERIAC

This plant is being grown more and more, probably because it has been on the television quite a few times. It looks rather like a nightmare head from the cinema, all lumps and bumps. It is also known as celery root – this old name was used in Jane Austin's *Emma*.

Celeriac has a wonderful celery flavour but with a harder texture, making a number of combinations possible in the kitchen that you just can't get with celery, particularly with meat. It is also hardier than celery, and can be kept in the ground

Growing for storage

Sow seeds in a propagator or in modules in March–April and thin growth to one plant per pot. Harden the young plants off before planting out in May into rows, with each plant 38 cm apart. The plants need plenty of circulating air to ensure they establish.

To ensure the soil is moist, apply a mulch around the plants and water well once a week. An addition of organic fertiliser will improve any pale crops. Remove any yellowing leaves to prevent knobbly stems. Overwintered crops should be protected with a layer of straw.

Harvesting

This plant is ready when the stems measure 8–13 cm in diameter. Trim excess roots and wash. The leaves are excellent in soups.

How to store

NATURALLY

You can keep celeriac in the ground for quite a while – but you run the risk of hungry slugs. It will keep for about 6 weeks stored in dry sand. Layer silver or river sand and celeriac in a wooden box and store in dry, cool conditions.

FREEZING

You can freeze celeriac. Although it is the devil's own job to cut celeriac raw, cut it into 1 cm cubes and blanch in boiling water for 90 seconds. Then plunge into iced water, towel dry and freeze quickly.

A number of recipes for celeriac involve using large slices of the root as a bed for steak. You can blanch and freeze slices of celeriac, but try to keep the slices thin, say 5 mm. In this case, blanch for only 1 minute.

CELERY

If ever there was a plant that needed lots of nutrients, it is celery. The amazing flavours produced in the stems are half condiment, half savoury and half dessert! Celery is an amazing plant, but goodness, it's hungry!

Care of the plant

They need good soil with lots of manure dug in, and frequently I grow them on a bed of manure – well rotted, of course.

Growing for storage

In order to keep celery for as long as possible, it should never experience water stress. It needs good drainage to avoid fungal problems.

Harvesting

Simply cut off stems you need for the kitchen. If you want to store them, choose an excellent looking plant and dig up the whole lot.

How to store

NATURALLY

They will keep on the shelf for about three weeks. Remove the leaves – they only cause evaporation and render the stalks flaccid.

FREEZING

Cut into ½ cm slices, blanch for 45 seconds in boiling water and then plunge into iced water. Towel dry and freeze quickly for up to a year.

DRYING

This plant has wonderful properties and you can dry it to make your own medicines – particularly for urinary problems. Drying is a matter of blanching and then towelling dry. The slices can then be placed in a desiccator until quite dry. Store in a dry jar on top of a layer of dry rice and it will last basically forever.

Celery Salt

Celery

Salt

- This is easy! Dry your celery as above. The celery has to be dry – really dry, so it snaps rather than bends.

- Then put the pieces in your food processor and whizz until they are powdered. Any big bits can be taken out and popped back into the dehydrator and then re-whizzed.

- Simply add equal volumes of celery powder and salt, mix well and store in a sterilised, dry jar.

COURGETTE

These are more or less seedless marrows that we only ever allow to grow to about 10 cm long. They are really lovely plants and you can eat the flowers too – usually in a stuffed form.

Growing for storage

They need plenty of water, but not splashed all over the leaves, especially on hot days as they succumb to fungal infections. Remove the flowers as soon as the fruits develop – otherwise you can get blossom end rot which affects the fruit, making the ends squidgy and rotten.

Harvesting

Simply twist off the fruits when they reach the right size. You need to keep your eye on them because they can grow rapidly in a short space of time.

How to store

NATURALLY

On the shelf they will last only a few days, a week at best. They will become floppy and therefore not too good to eat. Their skins are not very resistant to knocks and bruise easily.

FREEZING

Fortunately courgettes are good freezers. Cut into 1 cm slices and blanch for 1 minute in boiling water. Then plunge into iced water, towel dry and freeze quickly.

Courgette Soup

It is possible to make a batch of vegetable soup with courgettes as a base, or try this soup dedicated to courgettes. Freeze the soup once it has cooled. Alternatively, bottle and store for 3 months, being sure the bottles are properly sealed and sterilised.

Serves 4–6

Olive oil

1 onion, sliced

3 cloves garlic, chopped

1 kg courgette, chopped into small cubes

25 g butter

500 ml vegetable stock

Seasoning

- Lightly cook the onions in a little oil and add the chopped garlic.
- Add the courgette and, on a reduced heat, allow it to sweat a little.
- Add the butter and then the stock.
- Bring to the boil, then simmer for 30 minutes.
- Season to taste. Allow to cool before freezing for up to a year.

CUCUMBER

The love affair with the cucumber continues apace in greenhouses and polytunnels. They are easier to grow these days than in our parents' time, with extra straight fruits and smooth-skinned varieties that simply hang beautifully.

Growing for storage

Incorporate a large amount of compost well in advance of transplanting seedlings, and mulch throughout the growing period. Use raised beds in very heavy soils – they do not like cold roots. Indeed, they do not like anything cold, so avoid heavy soils with cold subterranean water.

Make sure they do not suffer water stress. Keep the watering regular and even. Remove the flowers, as they are self-fertile and pollinated cucumbers are more bitter.

Harvesting

Ensure you harvest cucumbers before you see any yellowing of fruit or leaf, as the fruit reaches 3–5 cm in diameter, from September onwards for outdoor grown cucumbers. Greenhouse cucumbers are ready from July. Cucumbers for pickling are best taken when they are 15 cm long, leaving table cucumbers to grow longer.

Use secateurs to cut through the stalk, and hold the cucumber as you cut it.

How to store

NATURALLY

They will last for a week in the fridge, or the coldest part of the kitchen. However, if you put a little olive oil on your hands and lightly rub this onto the skin of the cucumber, they will last another week – but you will have to peel them before use.

PICKLING

Put cucumbers destined for pickling in iced water for a couple of hours before salting. You get a crisper pickle that way.

Pickled Cucumber

Makes about 2 × 500 g jars

Pickling vinegar (see the Pickled Shallots recipe p.49)

250 g sugar

1 kg cucumber, sliced 5 mm thick

1 onion, roughly chopped

Salt

- Add the sugar to the pickling vinegar – you might need to heat the vinegar to get the sugar to dissolve.

- Place the cucumber and onion in a colander over a bowl and liberally cover with salt. Leave for a couple of hours for the water to drain out of the vegetables.

- Sterilise the jars and lids.

- Wash the remaining salt from the vegetables and dry them, before adding to the jars and pouring the vinegar over the mixture. Firm down the lids. Some recipes call for the pickle to be to be boiled in the jars for 10 minutes. Stores for a year.

FENNEL

Becoming more popular, this aniseed-flavoured vegetable is lovely steamed. The plant has wonderful medicinal properties, particularly for the digestive system, and it is really worth a small corner of the garden – if not for the beauty of the plant.

Growing for storage

Fennel has a tendency to bolt, so prevent this by sowing in modules in April and planting out into a cold frame or under a cloche when seedlings have four leaves. For germination, seeds must be at 15°C. Fertile soil, well drained and preferably sandy should be prepared with manure a season in advance. Keep the watering even – don't over-water and don't let them dry out. Old gardeners used to say 'They don't like wet feet!'

Do not plant in soil that has had lettuce or radish growing in it recently as there is a higher chance of fungal infections.

Harvesting

Harvest in July at soil level when bulbs are large. Leaving a stump in the ground will provide you with shoots that are good in salads. Cut them off at the stem and lift with a trowel.

How to store

NATURALLY
Fennel is not an easy plant to store naturally. It will keep for about 2 weeks wrapped in kitchen paper in the fridge. I have vacuum sealed it in the past and it has lasted around 3 weeks.

FREEZING
Cut the bulbs/stems into quarters. Blanch for 1 minute in boiling water, plunge into iced water, dry and then freeze quickly. They will last for a year.

PRESERVING IN OIL
Blanch and dry as above. Then transfer to a sterile jar and cover with olive oil, which will keep the fennel free from germs for about 6 months.

GARLIC

This plant, like all the allium family, not only tastes great, it has some fantastic healing properties and should be grown in every garden.

Growing for storage

Despite what they say, you can plant garlic corms at any time of the year. You get the best growth and return from those planted between November and Christmas. You are also best buying garlic designed for growing in the UK climate. If you plant supermarket-bought garlic you will get corms, but they will be small.

A well-drained, light soil should be prepared with a large amount of organic compost well in advance. Choose a sunny spot without too much shelter. Keep it well drained but never without water. If you can smell garlic while the plant is growing then it is likely to be too wet.

Harvesting

Garlic planted in October is ready in May and June, when the leaves begin to discolour. Clean and dry the bulbs, leaving them out in the open to cure over a few warm, rain-free days. Lever them from the ground with a trowel rather than just pulling them up. Cut off any leaves to reduce evaporation.

How to store

NATURALLY

I am not a fan of tying bulbs together 'French style' because in the UK at least it is often too humid and garlic can fall prey to fungal infections. Unless your storage area is perfectly dry, garlic will eventually succumb to fungal infections and, consequently, when you come to open the corms there will be little there to use. I prefer to keep it open on the shelf with a good air gap between bulbs, more the better. Garlic will last 4–6 months stored in this way.

As the chemicals in the corm continue to strengthen as garlic ages, it gets stronger as it continues to dry out over the course of a year.

DRYING

Rapid drying of sliced garlic is an excellent way of keeping the 'strength' of the garlic the same. Using a drying machine, you will have dry garlic in about 8 hours, to be kept in a clean jar on the shelf.

Peel the garlic and blanch the corms in boiling water for 45 seconds. Plunge into iced water and then slice thinly. Lay these in the drying machine and then dry on the coolest setting. A teaspoon of dried garlic is equivalent to about a clove.

FREEZING

Peel the garlic and blanch the corms in boiling water for 45 seconds. Plunge into iced water and then place the drained corms into a food processor. For every 500 g of garlic cloves, add 1 level teaspoon of salt.

Whizz the cloves until you have a paste. Transfer this paste into a sterile ice cube tray and cover with cling film. Freeze as quickly as you can. Pop out a cube of garlic as you need it – each ice cube compartment gives you two servings of garlic purée, depending on how garlicky you want your food to be. Frozen garlic will last for a year.

OIL STORAGE

Peel and blanch as described above. Add the corms to a sterile jar and then fill with oil to preserve them from infection. This will store for around 4–6 months.

Pickled Garlic

Making pickled garlic can be a little hit and miss, and it has to keep in the jar for some time before the garlic is ready.

Makes 3–4 × 500 ml jars

30–40 garlic cloves

Salt

Pickling vinegar (see the Pickled Shallots recipe p.49)

- Peel the cloves and place them in a colander over a bowl. Sprinkle over a liberal amount of salt and then leave for a couple of hours.

- Then wash the cloves and dry.

- Fill sterile jars with cloves and cover with the pickling vinegar.

- Leave the pickle to soak for a month before using. It is a bit of an acquired taste, but eventually you get addicted to it. Stores for a year.

KALE

Kale is a wonderful plant – it will grow anywhere, doesn't mind frost that much and will grow quite tall. It will grow from the beginning of spring to the onset of winter.

Growing for storage

The plant should be mulched and readily watered and perhaps fed with organic fertiliser in the early summer. Soil needs to be rich with good drainage to avoid water logging. Sow in modules in April–July and plant out two months later. If growing full-sized kale, ensure plants are spaced at least 60 cm apart. Plant dwarf kale about 40 cm apart.

To keep them well into the late autumn, cover with horticultural fleece.

Harvesting

Take leaves as they appear to be ready. Since you can have this plant fresh almost all the year round there is little reason for preserving it.

How to store

NATURALLY

Being leaves, this crop does not store that well – you will be able to keep it in a cool place for no more than a few days.

VACUUM SEALING

You can make it last up to a month in a sealed bag, but make sure you cook it well afterwards – there is no knowing what might be growing inside the plant leaf and stem.

FREEZING

Blanch in boiling water for 1 minute and plunge into iced water. Then compress into as small a space as you can and freeze. Again, cook well once you have defrosted it. Frozen kale can be kept for a year.

KOHLRABI

This is a brilliantly-shaped member of the brassica family. You are actually eating a swollen stem. It is an easy, handsome crop to grow, sowing in April and harvesting in July when the stems are only about 5 cm across – much bigger and they can become woody.

Growing for storage

So long as the soil is alkaline, this crop grows well. Keep it well watered, to avoid woodiness more than anything else, and you will have a decent crop.

Harvesting

Simply lift out of the ground when about 5 cm across and remove the leaves, as well as those leaves near the surface of the stem.

How to store

FREEZING

The usual routine: cut into quarters, blanch for 90 seconds and then plunge into iced water. Drain and pack into a freezer bag. Freeze as quickly as you can and they will last a year.

LEEK

Of all the vegetables in the world, the leek holds my admiration, not just for the flavour but the way it grows. To my mind it is the pluckiest of all the garden vegetables, the scrum half of the plot. Perhaps it's the way it stands up, that we top and tail it and simply plonk it into a wet hole in the ground. It fends for itself and rewards the grower with wonderfully versatile produce.

Growing for storage

The soil needs to be rich and well prepared beforehand. Dig in plenty of good compost. It needs to be free draining, in good heart: the kind of soil you would grow potatoes in.

Water only in periods of drought and feed with a nitrogen-rich fertiliser a month before harvesting. A regular routine of weeding is a must to prevent infection and to ensure a good crop. Buy the self-blanching varieties, and make sure they are not buffeted by strong winds in winter.

Harvesting

Harvest leeks when they are at least 2.5 cm thick, or as needed. Lever them out, rather than pulling them out, and wash off the soil as soon as possible.

They can remain in the ground, providing they are kept well weeded, and they are usually quite frost hardy.

How to store

NATURALLY

Leeks don't store well. They go slimy easily, but thankfully they are quite hardy. Leave them in the ground and they will last as long as you need them, or at least until you simply have to use that part of the plot again.

You can, at a pinch, wrap them in tissue and keep them in the fridge, but they don't keep for more than a couple of weeks.

FREEZING

This, apart from cooking in sauces and freezing, is the only way to store leeks. Cut them into rings and blanch in boiling water for 1 minute. Plunge into iced water and then freeze in bags of about 500 g, which is more or less enough for two servings. Stores for a year.

MARROW

Marrows and squashes of all kinds are wonderful to grow in the garden. For a start, they are easy to grow without any real problems, save fungal infections and the odd aphid attack with associated difficulties. They are such grand looking plants, the garden looks better for them too.

Growing for storage

They are hungry, thirsty, sun-loving plants, which need good soil with plenty of well rotted manure dug into it. Keep them well watered, in such a way that they don't get wet leaves. Mulch with rich compost to keep the moisture in.

I usually bury a pot in the ground next to the plant and water just in the pot. This keeps the humidity near the leaves as low as possible.

Marrows are frost intolerant. Sow indoors at 13°C in April and transplant in June into their growing positions under a cloche. This way you will get a spurt of growth. You can sow outdoors in June too, which means the plant will bear fruit later in August and into September. Sowing twice will mean you have fruit from July through to the onset of frosts.

Harvesting

Removing young fruit will encourage further fruiting. On second sowings you can allow some of them to grow to full size. Plants grown indoors develop monster proportions.

How to store

NATURALLY

Marrows need their own space and, since they are full of water and susceptible to fungal infections, they do not store well. You can wipe them with some antibacterial cleaner and dry them, standing them in their own space. They must not touch each other or any other veg for that matter, and they will last for about 6 weeks, maximum.

Often the shelf is a source of infection, and so some people store them in little nets hung from the roof, though I have never tried this myself.

FREEZING

You can freeze them, but don't blanch as they fall apart so easily if you do. Just cut them into cubes of approximately 3 cm and freeze them in a bag. They will last a year, but be careful to cook them very soon after defrosting.

MUSHROOMS

A lot of people are growing mushrooms these days. I suppose the joy is having them fresh, which is a completely different culinary experience than having them from the shops. You can also grow mushrooms that are difficult to buy in the shops, Jew's ear and shiitake being two excellent examples.

It is also possible to collect mushrooms from the wild. It is worth saying that of all the wild mushrooms we find in fields, woods and gardens, only 2% are poisonous and only 2% are edible. If you learn to identify the poisonous ones, Destroying Angel and Death Cap being the most obvious, and then the edible ones, Cep, the Edible Boletus, the Field Mushroom, the Chanterelle and the Shaggy Ink Cap, you can have wild mushrooms all the year.

Being fruiting bodies, mushrooms are not designed to last and most of them only have a distinct season. However, with a couple of bales of straw, and especially if you can enrich it with very well rotted chicken manure (it has to be extremely well rotted in a hot compost heap to kill the bacteria, especially salmonella), you can have fresh mushrooms all the summer.

How to store

NATURALLY
They don't last more than a week at best. Pop them in a paper bag and keep them in the fridge, away from other foods because they absorb aromas and will taint. You can put the paper bag into a polythene bag, which doesn't appreciably increase the keeping time, but does help with the intensity of absorption of aromas a little.

DRYING
It is best to dry mushrooms quickly, not naturally but in a desiccator. Use the lowest setting and it should take about 10 hours. You can slice the mushrooms to shorten the time and then store them in a jar on some dry rice. As long as they are perfectly dry, they will store forever.

FREEZING
It really isn't possible to store mushrooms intact. They pretty much have to be cooked and then stored. Otherwise they fall apart when defrosted.

Blanching them for 30 seconds in boiling water and plunging into iced water before patting dry and stuffing into a freezer bag does produce a frozen mushroom that when defrosted is good for mushroom soup.

However, another way to freeze them is to sauté them in a little butter, leaving them to cool and then freezing in a suitable container for up to about 6 months. It is possible to freeze them in vacuum-sealed bags, an individual portion at a time, for up to a year. Then all you have to do is place the bag in boiling water to heat through completely, and serve. (It is important that the contents have reached a minimum of 75°C, so give them a really good cooking.)

ONION

This information also applies to shallots.

There are two ways to grow onions: from seeds and from heat-treated onions called 'sets'. There are also two times for growing them. Japanese onions are sown in August or planted in September. They overwinter and are available from June to July. They do not store well at all and must be eaten fresh. However, you have the main crop to rely on fairly shortly after.

Growing for storage

The soil should be alkaline and well draining. Add a layer of organic compost before planting and dig over well. If the plot is prone to freezing, warm it with a layer of fleece before planting. A great tip for growing excellent onions is to plant them where you have previously had a fire. The wood ash makes the soil alkaline and the fire removes many pests.

Onion beds need to be well weeded and you can buy really good hoes just for the job. As the plants become established, stop watering to prevent them rotting.

Harvesting

Usually the plants are ready for harvest when the outer skins of the onions begin to brown and the leaves start to fall over. Lever them out with a trowel, rather than pulling by the leaves. Remove the leaves with a sharp knife and leave the bulb on a tray to cure. The outer leaves thicken and protect the fresh bulb within.

It is important you handle them correctly – don't throw them about and don't drop them. Any bruising will lead to rot.

How to store

NATURALLY

To be sure of good storage, keep them as dry as possible, though not so dry that they simply dry out. Any moisture will cause damaged tissue to rot. I am not a fan of tying them together unless you can be sure that any one of them will not become infected. I keep them on an open tray where they can each have air space around them. Equally, I don't use a sack for storage. In these conditions, you can have onions for a good three to six months.

A neat solution is to use tights or stockings. Pop an onion into the leg and tie a good knot in the leg over the onion. Then pop in another onion, tying off again. Continue until the leg is full. Each onion now has its own space and when you need an onion, simply cut it free from the bottom.

PICKLING

Obviously, pickling is a brilliant method of storing onions – although I actually prefer to pickle shallots. (See the recipe given in Chapter 5.) I usually make a batch of pickled shallots in September and they are wonderful by Christmas, and even better by Easter – if only I could resist them for that long.

FREEZING

I like to cut onion rings about 5 mm thick. Then chop them into quarters and freeze in freezer bags without blanching. They will store for a year.

DRYING

Use a dehydrator to dry onions. Before doing so, sprinkle salt over the peeled and chopped onions and leave them for a couple of hours. Rinse in cold water and pat dry, then dehydrate on the lowest setting. My machine gives me about 1 kg of dried onions at a time, which can then be stored in jars. Add the onions to soups etc. at an equivalence of 2 teaspoons per fresh onion.

Red Onion Marmalade

This is the easiest marmalade. You can pep it with garlic and herbs if you like but, to be honest, I think this simple recipe is the best.

Makes 5 × 500 g jars

2 kg red onions, peeled and roughly chopped

150 g butter

200 g sugar

350 ml sherry vinegar, red wine vinegar or just plain white vinegar

- Place the onion in a frying pan with the butter on a low heat. Cover the onions in the butter and then add the sugar while the onions are beginning to cook.

- Add the vinegar and cook very slowly, simmering away for about 30 minutes. You know the consistency is right when a spoon drawn over the surface of the pan gives a clear streak.

- Place in sterile jars and keep refrigerated. The marmalade should keep for 3–4 months.

PARSNIP

This aromatic plant is in the ground for longer than any other vegetable. You can sow it in April, and it isn't really harvested until the following February.

Growing for storage

The key to parsnips is two-fold: knocks and insect drilling make the root go mushy, brown and inedible. So, don't hit or damage this plant – always treat it with great care. And protect it from carrot fly – I use horticultural fleece to be sure of insect-free plants.

Make sure the soil is not stony, but well dug and slightly sandy. Parsnips need even watering – not a lot of water. Normal rainy months are fine, but top it up with a sparing splash of water if it hasn't rained for a few days.

Harvesting

It is purely a matter of taste when to lift parsnips. If you have a decent sized root in October, then use it. Generally speaking, they are available from November and right through the winter. A frost really sweetens the root; they do taste better after a hard frost.

Lever them out of the ground and then remove the leaves. Leave to cure in the air for a day or so.

How to store

NATURALLY

You can have parsnips from November to May (May is the latest I have had them, but it was a cold spring) just by leaving them in the ground. As long as they are actively growing they have reasonable immunity and stay fresh.

They will keep for about four months either in a clamp in the ground or in a box of dry sand (see Chapter 4 for information about these methods).

Approaching spring, the roots begin to grow again and they branch and soften. So you have to weigh this against the need to store for long periods – personally, I'd rather wait for new parsnips.

FREEZING

You can peel parsnips and blanch large chunks in boiling water for about 2 minutes; 1 minute for the thin end pieces of the root. Then plunge into iced water and freeze in freezer bags for up to a year.

PEAS

Peas burst into popularity in the seventeenth and eighteenth centuries. They were so popular that there were many stories of field workers picking and eating so many raw peas that harvests were severely depleted.

Growing for storage

Sow in good soil, with plenty of compost dug in. The key to peas is watering. They need plenty of water, but not so they are soaked through. A regular, daily dribble is sufficient.

Sow first early varieties in March and then successively. Main crop can be sown as late as June. You will then get a crop from June right through to October. They like good sun, but not too much heat – keep them out of the polytunnel and greenhouse. March sowings should be protected with a cloche.

Of course, garden peas (as opposed to field varieties) need support, and a trellis is ideal for this. The stronger the support the better – they don't like being blown about.

Harvesting

You can take peas as soon as they are ready, with more frequent pickings encouraging further growth. Of course, you need a good number of pollinators to work on your crop for best results.

You will see the pods fill. Take them when they are not quite full – you'll soon learn to recognise this stage if you grow peas. Rinse them under cold water to cool them before podding, which can be a few days later if necessary, but is best on the day of picking. The longer you wait, the less flavour the peas have.

How to store

NATURALLY

They will stay fresh for a week or so in the fridge. So if you are growing peas, there is really no point in storing them – they lose flavour and nutritional value.

DRYING

I have done this with the variety Kelvedon Wonder. Simply pod your peas and blanch in boiling water for 30 seconds. Remove to iced water and then leave to dry for a few minutes. Transfer to the dehydrator and dry at the lowest setting for about 8 hours. Make sure all the peas are really quite dry and hard before storing and they will last basically forever.

FREEZING

This is by far the best method to store peas for any time. Believe it or not, frozen peas were a Victorian invention, though it was a long time before they became popular. Blanch for 30 seconds in boiling water, plunge into iced water and dry. Pack into freezer bags and freeze rapidly. They will store for a year.

PEPPER

These are becoming more popular and several easy-grow varieties are available. They are hothouse plants, often needing temperatures over 20°C for germination. They range from the mild capsicum to the medium jalapenos to the seriously hot Scotch Bonnet. Actually, I cannot deal with really hot ones myself, but there is no accounting for taste!

The older a pepper gets, the hotter it becomes. As a general rule they all start out green and turn red as they mature. Another rule of thumb is the smaller the pepper, the hotter it is. If your peppers have aged so much that the skin is beginning to peel off, leaving whitish lines on the surface, then it is going to be very hot indeed.

Growing for storage

The best fruits come from plants that are grown in rich but light compost – I add a little sand to my compost as it helps keep it light. They are water, temperature and nutrient hungry plants. I treat them like tomatoes in this respect, watering almost daily (certainly in July and August) and providing them with plenty of air circulation to keep infections at bay. I feed them weekly with a high potash tomato feed.

Harvesting

Cut peppers away when they are big enough to be used. With hot varieties I wear plastic gloves to keep the chemicals off my hands, which otherwise can have disastrous results! Cut a little of the stalk along with the pepper. Taking young ones early as they mature on the plant will encourage new growth. Use the second fruits for storage for capsicums and long peppers. Hot peppers can be taken when there are a lot on the plant, allowing you to store them all together.

How to store

NATURALLY

Capsicums will last a month in the fridge or on a shelf. They do better if wrapped in a sheet of damp kitchen paper. Hotter chillies can be left on a shelf in a tray where they will slowly dry, and get hotter in the process. Like this, they will keep for 2–4 months.

FREEZING

Peppers do not need to be blanched, but I like to prepare them before freezing. Wash and slit them lengthwise into two pieces and then de-seed. Pack into freezer bags and freeze quickly. They will be good frozen for up to a year. Small peppers can be frozen whole. Defrosted they are slightly softer than fresh chillies, but are quite acceptable for cooking.

DRYING

Smaller, hotter chillies are best dried and there are a number of ways of doing this. You can simply leave them on the plant, stop watering it, and as the leaves fall off the peppers will be drying out. When they become leathery and dry to the touch, pick them off the remains of the plant and put them in a jar on a bed of dry rice. They will last all year.

You can pick peppers when they are just red and leave them on a tray in a dry place, covered with kitchen paper to dry out naturally. Store as above when they become leathery.

You can dry them completely in a dessicator, which will take about 10 hours, and then grind them in a coffee grinder (remove the stalks) and make some hot chilli powder.

Jalapeno Pickle

These are amazing, and I have to confess that if I open a bottle, I eat the lot. I simply cannot resist them. It is quite salty and because of the water it is milder than other pickles.

Makes approx. 2 × 500 g jars

Approx. 15–20 jalapenos

1 litre pickling vinegar (see the Pickled Shallots recipe p.49)

500 ml water

Juice of 1 lemon

50 g cooking salt

- Pick the jalapenos when they are green – if they are underripe they keep their crispness. Cut them in half and remove the seeds.

- Pack them into jars (Kilner jars are best) but not packed so tightly that they would be difficult to remove.

- Add the pickling vinegar to the other ingredients. Pour over the jalapenos and seal the lids.

- Because the pickle is weakened by the water, it needs to be sterilised. Place them in boiling water to the neck and boil for 15 minutes, then leave them to cool and they will be fine for a year, though I have never been able to keep them for a year myself!

POTATO

Growing potatoes is possibly one of the most rewarding experiences you can have. The flavour of just dug potatoes is wonderful, a complete reward for your efforts.

Growing for storage

It is important that potatoes do not see sunlight, so earth them up – that is, cover them with earth to keep them in the dark. Make sure they never dry out, but equally that they are never too wet. Also make sure that the soil is well prepared before planting with as much well rotted manure worked into the soil as you can manage. This should be done a good 6 months before planting.

Plant seed tubers in a drill 15–20 cm deep, with any sprouting points facing upwards. Leave 45 cm between tubers. Remove flowers as they appear and take measures against blight. Never water the leaves, only the soil; a seeper hose is a good idea. If others nearby have blight, remove the vines and prepare to store.

Harvesting

There are three or four varieties of potatoes within the broad groups of earlies and maincrop. These take progressively longer to mature, from 12 to 15 weeks. It is therefore possible to have potatoes to harvest from June to October and beyond. You can leave potatoes in the ground until the first proper frosts.

A fortnight before the potatoes are ready for harvest, cut down the foliage growth so stems are only 10 cm above the level of the soil. This will improve the skins on the tubers. Using a fork, dig up the plant and carefully separate the potatoes from the root. Leave the potatoes in the air to cure for three days – not long enough for them to be bothered by the light, but enough to thicken their skins. Then store.

How to store

NATURALLY

Potatoes can stay in the ground until the first harsh frost, and consequently they do well in a clamp which, because of its extra insulation, is a good store right through the winter.

I prefer to store potatoes in a shallow but wide box indoors, so there is plenty of air to circulate around the potatoes. I don't wash the potatoes before storage, but I do rub off some of the thick pieces of clay from my heavy soil.

From time to time, as you collect potatoes for the kitchen, have a good inspection of the potatoes in your storage box and discard any that look as though they might be going off. This is the reason why I don't keep them in a sack: a bad potato somewhere deep inside can ruin a bag of potatoes, and you will not easily find it until it is too late.

PUMPKIN

Great fun: pumpkins both in soup and as a lantern. The warmer the better for pumpkins, as is a fertile soil with good drainage and an open, sunny aspect.

Growing for storage

Warm the soil with fleece or plastic before planting out, and sow indoors in April – May for best results. You can sow in June outdoors, but you do not get long enough of a season to get good fruit.

Pumpkins root deeply and therefore do not need as much water as other cucurbits, so simply add a mulch to retain the water in the soil. Add support to heavy fruits to prevent stems from snapping, and add netting around individual fruits to prevent them sitting on the soil.

Harvesting

Pumpkins are ready when they have fully developed their colour and sound hollow when tapped. Cut them from the stem leaving about 10 cm of stalk attached. The skins of cut fruit will harden in the sun, so leave them to cure for a couple of warm days.

How to store

I have found that the best way to store pumpkins is to cure them well and handle them carefully. They will develop a hard skin that is difficult to mark with the fingernail. Then all you have to do is take them to a cool, dark, dry place to store for about a month. Do not store them near other fruit as they will deteriorate due to the ethylene gas given off by ripening fruit.

RADISH

Radishes are growing in popularity and they are completely easy grow plants. You can find a radish to harvest from May to November, and you can grow them indoors too, so there shouldn't be a month when they are not available.

Growing for storage

A rich soil with good moisture retention is necessary, preferably a warm soil. Sowing in cold months should be preceded by covering with a sheet of black plastic to warm the soil. Add a layer of organic compost well in advance of sowing.

If you let radishes grow too long they become woody and fibrous and less pleasant to eat. Sow seed every two weeks to ensure a crop throughout the season. Sow every two weeks from March until September, to provide crops until December.

Harvesting

Summer radishes should be harvested as soon as the roots are a decent size for eating. Winter radishes can be kept in the ground until needed, but protect them from frosts with a cloche.

How to store

NATURALLY

Radishes harvested in the summer will last a week in the fridge. Remove the leaves. If they need to be kept for longer they will go soft, but they can be rejuvenated by a couple of hours soaking in iced water.

They also store well in dry sand, where they will last about six weeks. Don't wash them until you need to use them, and perk them up with cold water as described if necessary.

SALADS

The following advice applies to all salad leaves: lettuce, endive, beets, rocket, dandelion – almost anything composed of thin, delicate leaves.

Growing for storage

Salads can be grown all the year round, especially if you have the opportunity to grow under cover. October sowings of most varieties, grown under cover in a greenhouse or even a sealed cloche, will be edible through the winter. If you experience really bad frosts, a night light candle is all you need to keep the cloche frost-free.

Rocket and dandelion (I would love to see people harvesting dandelion in spring like we did a hundred years ago – it is such a health-giving plant) are more delicate and need to be sown in spring. That said, successive sowings will give you a crop until the end of September at least.

Similarly, chards and beets grown for their leaves are good from May to September, if sown from April to the end of July/early August.

Any plant with large leaves does not do well being transplanted. For winter-grown lettuce, sow in pots and pick out smaller plants, leaving one pot per plant to grow on. Spring and summer sowings should be done in the place where they are to grow and mature.

If you do sow in modules or pots, plant out as soon as they have four leaves, with the soil just beneath the bottom two leaves. Keep well watered until the plants establish.

Harvesting

Salad leaves can be gathered as soon as they appear ready – try to take them as young as possible. Leafy lettuces can be cut and grow and left for secondary growth. Hearting lettuce should be fully cut to ensure they do not run to seed.

How to store

NATURALLY

As we have already surmised, salads should be planned for maximising fresh stock, pulling them from the ground as they are required. If you have to store them, wrap them whole or in single leaves in damp kitchen paper. This way they will remain fresh for not much more than ten days.

In February the first dandelion leaves are appearing, and they do taste good, so long you can guarantee the dog or cat hasn't been on them, which is why you should cultivate your own!

SPINACH

This vegetable has a reputation of being difficult to grow, but it is not that bad really if you give it a fertile soil that is moisture retentive. Don't overfeed spinach; it concentrates nitrates in its leaves to levels that can be quite dangerous if overfed.

Growing for storage

The answer for great spinach is even, constant water and space. They need arm room to do really well. The larger the space, the better they are, so give them about 40 cm between plants. Sow closer together and then thin them out. Water consistently through dry periods.

Harvesting

You can take leaves as you need them, or harvest everything when you want to, as soon as the leaves are worth taking.

How to store

NATURALLY
Being all leaves they don't keep very well, no more than a week really, wrapped in kitchen paper and refrigerated.

FREEZING
I cook a batch of shredded spinach in salty, boiling water for a couple of minutes. Allow to drain in a colander and then pack into bags of 500 g and freeze quickly. I use 500 g because that is a family portion (for us) but you might want a different quantity. Frozen spinach lasts for a year.

SWEDE

This slow growing plant will stay in the ground a long time, and stores well. It is very much worth keeping in the garden for as long as possible, and is seriously tasty roasted.

Growing for storage
You need a poorish soil, slightly acidic and manured at least two seasons previously. The key is water; every time the plant suffers water stress it compensates by becoming fibrous and hard. Don't overwater, but don't let it dry out. Weed the beds thoroughly, keeping on top of this throughout the growing period.

Harvesting
Lift swedes, that is lever them out with a fork, when the roots are 10–15 cm in diameter and remove the leaves.

How to store

NATURALLY
They will stay on a shelf for up to two months quite readily. They will dry out a little and become a little hard on the surface, but this can be peeled off before cooking.

You can clamp them, especially if they are to overwinter, when they will last for a good 3–4 months. They also do well in dry sand in a box. Try to store smaller swedes, leaving the mighty ones for immediate cooking.

FREEZING
Peel and cut your swede into 2 cm chunks. Blanch for 90 seconds in boiling water and then plunge into iced water. Dry and pop into freezer bags and freeze quickly. Stores for a year.

I sometimes freeze carrot and swede mixed and completely cooked through. I personally like them about 1 cm cubed, seasoned with salt and pepper. Having been drained and frozen, they only need bringing to the boil when you want to use them.

SWEETCORN

Sweetcorn is a type of grass that is pollinated by the wind and provides an impressive show when grown in the garden. It is best to plant supersweet varieties to ensure the best flavour and the secret to getting good crops is warmth and wind.

Growing for storage
Always grow your sweetcorn in a grid rather than a line to aid pollination. This is really important. If you grow them about 45 cm apart in a grid, then no matter which way the wind blows, you will get good pollination and therefore better crops.

Don't sow in cold soil. If you must sow them directly into their growing spaces, warm the soil with a black plastic sheet. Soil should be light and in a sunny aspect. If you cannot guarantee about 20°C, then sow indoors and transplant in June, and if you are in northern climes, think about growing sweetcorn indoors.

Harvesting

As flowers turn brown and appear desiccated, test the fruits for ripeness. Harvest by snapping off the cobs. They are ready when you pierce the seed with a fingernail and the juice inside is not watery, but creamy white.

How to store

NATURALLY

They only last about a month in the fridge or on the shelf. Keep the leaf around the head and keep air spaces around the cobs.

FREEZING

You can freeze in two ways. Blanch whole cobs in boiling water for 5 minutes, plunge into iced water and then dry and freeze in freezer bags. You can also use a knife (be careful and use a towel to hold the corn so you don't cut yourself) to remove the individual seeds, blanch them for a minute, pop into iced water and then dry and freeze in bags. Either way, frozen corn will store for a year.

DRYING

To be honest, I don't do this often. For a start, I don't have that much room for corn, and what crop I do get is eaten fresh. However, you can do it by removing the seeds, put the oven on its lowest possible setting, leave the door open and dry the corns on a tray. Then store them in a sterile dry jar ready for use. They are great in soups.

Corn Relish

Makes about 1.5 kg

400 g sweetcorn

1 small onion, finely chopped

1 small red pepper, finely chopped

180 g white caster sugar

200 ml white wine vinegar

½ teaspoon turmeric

1 level teaspoon mustard powder

1 teaspoon salt

1 rounded teaspoon cornflour mixed with 2–3 tsp of the vinegar to make a paste

- Cook the sweetcorn until it is ready to eat, then leave it to cool.

- Remove the corns from the husk with a fork and place in a pan with all the other ingredients. Stir over a low heat until the sugar has dissolved.

- Bring to the boil. Then simmer for 15–20 minutes or until the onion and pepper are tender and the relish is thick.

- Allow to cool completely before serving or immediately potting into sterilised jars. Stores for 6 months.

TOMATO

There is a tomato for every occasion and every garden. They are what summers were made for and are completely versatile too.

Growing for storage

The key for great storing tomatoes is warmth, water, nutrients and variety. If you want to make passata, choose Italian plum tomatoes like Pomodoro or Roma. Good old chutneys need the old favourites such as Alicante.

Sow in March indoors, in pots filled with organic compost, one to three seeds per pot, and thin to one plant. Ensure the temperature is higher than 18°C for germination to take place, either by heating a greenhouse or a specialist heated propagator.

Water well in the growing period, especially with plants that are growing in pots or containers, but do not overwater, as this can alter the flavour. When the flowers appear, feed weekly with tomato fertiliser.

Harvesting

Allow the fruits to ripen on the vine so the other fruits get the message too and ripen up.

How to store

NATURALLY

You really cannot do better than leaving them on the vine and taking them as you need them. They do not store well on their own – they will last a week or two in the fridge, that's all.

DRYING

Chapter 5 includes instructions for making dried tomatoes using a dryer. You can, assuming you can guarantee a fortnight of really hot weather, make sun-dried tomatoes, but put them under horticultural mesh to keep the birds and insects off.

CHUTNEYS

Obviously, relishes and chutneys are important methods of storing tomatoes. Chapter 5 includes recipes.

Ketchup

It is so important we allow children to eat ketchup, especially boys. It is so full of lycopene and a good, regular dose of lycopene provides excellent protection against testicular cancer, making it an excellent adjunct for developing boys. Ketchup is easy to make at home with a glut of tomatoes.

Makes about 1 litre
  500 ml white vinegar

  2 level teaspoons pickling spices

3 kg ripe tomatoes, sliced

1 large onion, finely chopped

2 Bramley apples, peeled and diced

180 g sugar

½ teaspoon paprika

3 teaspoons salt

- Boil the vinegar and pickling spices for 10 minutes and then allow to cool. If you wish, you can remove the spices after 10 minutes (having placed them in a muslin bag for easy removal) or leave them in for a spicier ketchup.

- Mix the tomatoes, onion and apple in a large pan and simmer for 30 minutes until everything has become soft and pulped.

- Force the pulp through a small sieve and return to the pan.

- Stir in the vinegar, sugar, paprika and salt and bring to a fast simmer until the mixture resembles double cream in consistency.

- Pour into sterile bottles or Kilner jars. Seal the lids and boil in a water bath for 10 minutes. Label and date when cool. Stores for 6 months.

TURNIP

This is my favourite vegetable for roasting. It builds some amazing flavours and I just love it!

Growing for storage

Turnip is an easy grow plant – you just sow it and off it goes. It prefers a cool, nitrogen-rich soil, slightly acidic with plenty of compost.

In February, under a cloche, sow thinly directly into final growing positions. From April to the beginning of June you can sow in the open. This way you get a harvest from June to September. You can also sow in September for an autumn crop, especially under a cloche. The key is to weed well and keep the water at a constant 'slightly damp'.

Harvesting

Lift with a trowel when the plants are about 8 cm across. Remove the leaves and tap root and leave outside for a day to cure.

How to store

There really is no substitute for keeping in the ground. They clamp well and will keep equally long in dry sand. You can keep turnips for about two months in this way.

CHAPTER 7

A–Z OF GROWING, STORING AND PRESERVING FRUIT

This chapter provides information on making all those wonderfully tasty fruit preserves as well as how to preserve fruit in its natural state for as long as possible.

By its very nature, fruit is meant to be eaten and consequently it simply does not last all that long without being processed. But, of course, there are exceptions to every rule.

Packed with sugar and water, it makes an ideal medium for the growth of spoiling organisms. So it is best processed as quickly as possible after collecting and, owing to its transient nature, it needs to be handled with far greater care than vegetables.

Always be careful on the journey from the garden to the kitchen, as it is then that fruits are most likely to be damaged. Tissue damage from dropping or banging will severely reduce or even wipe out any natural storage properties.

Whereas you don't always wash stored vegetables until you need to use them, always wash fruit as soon as you can, preferably in very cold water.

Many fruits take a while to get established, and the first three years of any fruit tree is an important time. Preparation is the key, from planting the tree right through to your latest harvest. Normally it is important to take the time to plant fruit properly. Add plenty of well rotted manure to the fill mix, and stake the tree securely. For the first three years of plants like apple, pear and plum, mulch in the spring with rich compost and feed in the height of the season with organic fertiliser. This good early start will bring you many years of excellent fruiting.

APPLES

Growing apples for storage can be daunting. There used to be as many apple varieties as there were parishes, but sadly many have long since disappeared. Moreover, the numbers of apples grown and stored in our kitchen gardens have depreciated. It is important that apples are grown with sufficient care or else they will spoil in the keeping.

Early years

Apples should not be allowed to harvest for the first three years. I usually remove all developing fruit with the pruning shears at the end of June. The logic to this is that cutting the fruit in June mimics the June drop of apples that occurs naturally, so taking all the fruit then will not be out of the ordinary for the tree.

The reason for taking the apples at this point, and composting or disposing of them, is to allow the tree to build its strength and resilience before its main work of growing and cropping. Trees that have fruit taken too early will take a lot longer to come into good production, and the fruit will never be as good as it could have been.

During these early years, and indeed, the rest of its life, give the tree a good mulch (half a barrow load is not too much) of 50:50 good compost and well rotted manure – not touching the bark, but in a circle around the base. Do this in the early spring. In June, July and August, give the tree a spray with diluted organic fertiliser.

Harvesting

There are many types of apples on the market, some which produce early fruit, for example mid summer fruit like Grenadier which can be taken in late July – early August. Others are known as mid season fruit that are pickable in September, such as Cox's Orange Pippin. Then there are late season types, such as Bramley's Seedling, which are harvested in October.

Consequently, it is possible to have fresh apples for at least four months of the year if you have a plot around 10 m by 10 m for standard trees, or a warm south-facing wall 10 m long for espaliers.

When is it time to pick? The tree makes a layer of wood, which dries out and becomes brittle. Eventually it stops bearing the weight of the apple and the apple falls. If you cup the apple and twist it, the apple should simply fall into your hand. Do not force the apple off the tree, and do not cut it.

How to store

Firstly, never store blemished apples, or ones with holes from insects or bruises from falling. Don't store apples that have fallen. Clean each apple with dry kitchen paper and then wrap it in more tissue. Place the apple on a shelf so that it does not touch any other apples or fruit.

Kept dry and in a reasonably cool place, a cellar is ideal so long as it is dry (I used an old fridge for many years), your apples should be fine for at least three months, probably nearer to five months.

It is only the mid to late season apples that are worth storing. Early apples are not that good at storing and go off in the first few weeks, but given that you can have more apples on the go by the end of the month following harvest, this doesn't matter.

PURÉE AND BUTTER
See Chapter 5 for recipes for making Apple Purée and Apple Butter.

Apple Jam

To vary the flavour, at the start of cooking the apples add a small piece of cinnamon stick or teaspoon of ground cinnamon. Remove the stick before adding the sugar.

Makes approx. 5 × 500 g jars. To make larger quantities, for every extra 500 g apples add an extra 500 g sugar, 200 ml water and the juice of ½ a lemon

Approx. 1.5 kg cooking apples (Bramleys are ideal)

600 ml water

Juice of 2 lemons

1.5 kg sugar

- Peel, core and dice the apples and place in a pan with the water and lemon juice. Put the apple peel and cores in a muslin bag and add to the pan. This helps the set – if you prefer a softer set then omit this.

- Bring to the boil, then simmer until the apples are soft and pulpy. Use tongs to lift out and squeeze the muslin bag. Discard the bag of peel.

- Add the sugar and stir over a medium heat until the sugar has dissolved.

- Bring to the boil and boil for 5 minutes. Then test for setting by drizzling a teaspoon onto a cold saucer. Allow the jam to cool a little and then, using your finger, push it. If it crinkles, you know your jam is ready.

- Allow to stand for 3 minutes, then ladle into sterilised jars. Secure the lids well. Label and date the jars when cool. Stores for about 4 months.

Apple Jelly

Bramleys are best but you can use any apples. If you are using sweet apples mix them with about 10% Bramleys, or even crab apples.

Makes approx. 5 × 450 g jars

2.5 kg cooking apples

Juice of 1 lemon

Water to just cover

- Cut the apples into small chunks – there is no need to peel or core the fruit. Follow the method for making fruit jelly described in Chapter 5.

Simple Apple Chutney

Use malt vinegar if you prefer a browner chutney with a slightly caramel flavour. White vinegar gives a lighter colour and milder taste. The type of sugar you use will also affect the chutney's colour and taste.

Makes approx. 8 × 500 g jars

1 kg onions, peeled and finely chopped

3 kg apples (any kind will do – a mixture of dessert and cooking gives an interesting flavour and texture to the finished chutney), peeled, cored and chopped into small pieces

25–30 g salt (depending on how salty you like your chutney)

20–25 g ground ginger

1½–2 level tablespoons ground cinnamon

1 litre malt or distilled white vinegar

150 ml water

1 kg soft brown or granulated white sugar

3 tablespoons honey

- Place the apples and onions in a large preserving pan and add all the other ingredients.

- Stir over a low heat until the sugar dissolves. Raise the heat and bring slowly to the boil.

- Reduce the heat and simmer for about 1 ½ hours or until thick and smooth. Stir regularly.

- Pot whilst still very hot in sterilised jars and seal. Allow the jars to cool before labelling.

- Allow this chutney to mature for at least 6 weeks before eating. It will store for about 4 months.

JUICE

It is easy to extract the juice from apples, either by the crush/press method as described in the Chapter 5 or by steam juice extraction. The pulp left behind from the crush method is brilliant made into a granola bar.

CIDER

Traditionally cider is made in a barrel, for which you need a good few trees to fill with juice. The apples are first crushed or scrumped up with a masher, and then pressed between sacking to extract the juice. This is then simply poured into the barrel and sealed. Wild yeasts on the skins of the apple ferment the sugars in the juice and, after a year, the cider is ready.

For the small producer, however, apples are crushed and pressed as per the extraction of juice in Chapter 5. The juice is then transferred to a sterile demijohn and a teaspoon of brewer's yeast added. This then converts the sugar to alcohol.

The cider needs to be racked off – that is siphoned so that the lees (all the dead yeast cells) are left behind. The cider is transferred in the process to a new sterile demijohn. At this point give the cider a good shake to remove most of the carbon dioxide and you will have clearer cider as a result. Leave to stand and rack again before bottling, or simply drinking from the demijohn as in our case.

APRICOTS

I don't know why we don't grow more apricots in this country. They originate in Eastern Europe and are mostly frost hardy, except when it flowers in the spring and is nipped by late frosts.

Growing for storage

The plant is fairly easy care and will fruit abundantly, especially if planted in full sun but in a protected spot. It particularly does well trained on a south-facing wall. It requires little in the way of pruning and needs a feed in the spring. The apricot doesn't suffer from the same diseases as peaches but, more than anything else, it stores quite well. Keep it reasonably well watered, and net to keep the birds off.

Harvesting

In the summer when the fruit's colour has ripened, they are to be found full and not too hard, nor too soft. Taste one and if they are sweet, they are ready. Most importantly, they are easily damaged, so be careful!

How to store

They will keep on the shelf for about a month, though the texture will become furry the longer you keep them. Wrap them in tissue and keep them in an open box and they will last for two months.

Apricot Jam

This is one of the easiest jams to make, as since apricots are quite full of pectin it sets very easily. Some recipes require water – you might want to add a small amount if the juice isn't forthcoming. Also, some recipes ask for one of the stones to be boiled with the jam for the pectin content, although I haven't found this to be necessary.

Makes 3–4 × 500 g jars

1 kg just ripe apricots

700 g granulated sugar

- Cut the apricots open, discard the stones and cut into quarters.
- Add to the pan with the sugar and heat slowly, stirring all the time.
- Once melted, bring to a rolling boil and stir continuously.
- Test for a setting point and repeat the boiling until you get one.
- Bottle into sterile jars. Stores for 4 months.

Apricots in Syrup

These are lovely and make a brilliant sweet.

Makes approx. 3 × 500 g jars

1 litre water

500 g sugar

1 kg apricots

- Boil the water and carefully add the sugar. Allow to cool and clear.
- De-stone and quarter your apricots.
- Add the apricots to sterilised jars and fill with syrup to the neck, making sure there are no air spaces. Seal and boil in a water bath for 15 minutes. They will store for about 4 months.

BLACKBERRIES

Blackberries can be found growing wild all over the place. They are considered a nuisance, but in truth they are a great bounty. Be careful to collect berries only from those plants for which you know the provenance of the soil. There are also a number of cultivars available for the gardener with stronger flavours and more stable fruits.

Blackberries are available from July to September. There is a saying that the devil urinates on blackberries after Michaelmas (29 September), so don't pick any after that!

Harvesting

It takes a long time to painstakingly remove the berries from the plants. Drop them into a clean food grade bucket and it is a day's work to collect about 8 kg.

If you want to make blackberry jam, you need to collect the fruits in smaller containers. If you simply fill a big bucket, the ones at the bottom will be crushed, releasing their juice. This will make it difficult to make jam because the pulp will be comparatively juiceless.

Blackberry and Apple Jam

The reason for adding apple to this jam is that apples contain a lot of pectin and add texture to the blackberries without detracting too much from the blackberry flavour. The earlier you pick your blackberries, the more pectin they have. A good set jam will be made from early blackberries.

Makes 10–12 × 500 g jars

1 kg apples

2 kg blackberries

200 ml water

Juice of 2 lemons

3 kg sugar

- Leave your apples unpeeled. Core and cut into 1 cm cubes.
- Place the fruit in a pan with the water and simmer gently for about 10 minutes until it is soft and the juice is running.
- Remove from the heat and add the lemon juice.
- Stir in the sugar and replace on a low heat. Stir gently but constantly until completely dissolved.

- Turn up the heat and bring to the boil. Boil for 4 minutes, checking for setting after this time.

- When setting point is reached, remove from the heat and cool for 2 or 3 minutes. Then stir to distribute the fruit evenly. Pour into sterile pots and seal. This jam should keep for 6–9 months unopened.

Blackberry Wine

Makes approx. 5–6 × 750 ml bottles

5 kg blackberries

500 g plain white sugar

Juice of 1 lemon

1 teaspoon wine yeast

- Place the berries in a muslin sheet and tie off. Pour a kettle of boiling water into a bucket with the muslin sheet of berries and then immediately mash the bag with a potato masher until all the juice has been removed.

- Put the juice into a large pan and bring to the boil, adding the sugar.

- Allow to cool and then decant into a demijohn. Add the lemon juice and wine yeast. Seal with an airlock.

- In a fortnight, rack off as per cider making: siphon the wine into a new sterile demijohn so that the lees are left behind and give it a good shake. Leave to stand for a week and then rack again before bottling into sterile bottles.

- In 6 months this wine will remind you of a summer like no other. It usually stores for a good year.

BLACKCURRANTS

The aroma of blackcurrants is first perceptible when you are working on the plot in early spring. A waft on the wind reflects the aroma of blackcurrant as the sap rises in the plant. From that moment on I need a drink of Ribena!

Growing for storage

These are hungry and thirsty plants. Mulch them in spring with good quality compost and also feed them as the flowers come into play. When the fruits are developing, don't let them do without water.

Harvesting

This is the easy bit: as soon as the fruits change from green to black, they are ready. Always pick them on a nice summer day when the fruits are dry – wet blackcurrants go furry with fungal growth very quickly indeed.

How to store

CORDIAL

Use a steam juicer to get the best of the juice from blackcurrants. This is the easiest way to get the juice out without resulting in pulpy mess. You can use a press, and strain the juice through a muslin or jam bag, but it doesn't work so well.

This cordial can be sweetened by adding 300 g sugar per litre, and then bottling and pasteurising. This gives you the basis for a great cordial that will last all year and that is easily diluted to make a refreshing drink.

JUICE

Simply crush and press. You can put the berries in a muslin bag and press them in a cheese or multipurpose press. The juice will freeze well for a year, but kept uncooked it will ferment quickly.

Blackcurrant Jam

Makes up to 10 × 500 g jars

2 kg blackcurrants, stalks removed

1.5 litres water

Juice of 1 lemon

2.5 kg sugar

- Add the blackcurrants and water to a large pan. Bring to the boil and simmer for about 40 minutes.

- Remove from the heat and add the lemon juice.

- Stir in the sugar and replace on a low heat. Stir gently but constantly until the sugar has completely dissolved.

- Turn up the heat and bring to the boil. Then boil for 4 minutes, checking for setting after this time. If necessary, repeat until you get a setting point.

- When setting point is reached, remove from the heat and cool for 2 or 3 minutes. Then stir to distribute the fruit evenly. Pour into sterilised pots and seal. Stores for 6 months.

FREEZING

Remove the green stalks etc. from the blackcurrants, wash and spread them out on a tray still damp. For every kilo of blackcurrants, sprinkle 1 heaped tablespoon of sugar all over the fruit and mix by hand until they are all covered with sugar. Fill a freezer bag or an ice cream tub with this mixture and freeze as quickly as possible. You can freeze individual portions – great on ice cream!

Blackcurrant Wine

It is not easy to make wine with fruit that is so full of pectin. Use a spoonful of pectolase, an enzyme that breaks down the pectin and releases the full fruit. You get more lees but this is discarded in racking.

Makes approx. 5 × 750 ml bottles

2 kg blackcurrants

2 litres apple juice (any fresh juice will do, even juice bought from the supermarket although be careful that it doesn't contain any preservatives)

500 g white sugar

1 teaspoon pectolase

1 teaspoon wine yeast

- Juice the blackcurrants and add to a pan with the apple juice. Bring to the boil and add the sugar. Allow to cool and then add a teaspoon of pectolase.

- Leave to stand for 2 hours and then decant into the demijohn.

- Add the yeast and an airlock. Put the demijohn in a large pan to catch any spillage, which often comes through the airlock.

- When it has stopped bubbling, rack off, wait a fortnight and rack again. Stores for 1 year.

CHERRIES

The glory of spring, the cherry gives the most amazing fruit in the world. The fruit's flavour is down to the cyanide ion, which makes you think.

Growing for storage

Planting cherries needs patience, good fertiliser and lots of sunlight. The flowers are food for bees early in the spring, so if you keep bees you are preserving cherry nectar without lifting a finger. Often cherries need a pollinator tree nearby because they are not self-fertile, so if you grow two trees you will be doubly blessed with the certainty of a good crop.

Harvesting

When the fruits are a bright, deep red they are ready to take. There are two types of cherries: cooking and eating. The cooking varieties are smaller and more acidic than those for eating.

How to store

FREEZING

Cherries do well frozen, with the stones in or out (I prefer to remove them). Stone them using a cherry pitter, which is essentially a spoon on a hinge with a hole in it and a prong opposite the hole to push the stone out. Wash and freeze, either with or without sugar. They will store frozen for a year.

CHERRY BRANDY

Don't mess about with cheap brandy, you always get a cheap result. Buy some dear stuff and have three brandies before you go to bed. Pop some stoned cherries (no, not stoned on brandy) in the bottle until the liquid rises to the level it was before you vacated it. Screw the lid tightly and hide the bottle away. Bottle in August and keep until Christmas, when your favourite aunt can have a tipple.

DRYING

Cut your cherries in half and place in the dryer for best results, on the lowest setting. Using an oven is just a little too harsh. The outdoors can work, but you need three days of wall-to-wall sunshine and a good bird net. When leathery and lusciously thick, they are at their best. Dried cherries will store forever as long as they are perfectly dry.

Cherries in Syrup

Makes 3–4 × 500 g jars

500 ml water

500 g sugar

1 kg cherries

- Boil the water and carefully add the sugar. Allow to cool and clear.

- Stone and wash the cherries and place in sterile jars.

- Pour over the cool syrup to the jar's neck, making sure there are no air spaces. Seal and then boil the jars in a water bath for 15 minutes. Stores for about 6 months.

Cherry and Apple Jam

There's not much pectin in cherries so I like to bulk them out with apple.

Makes 4–6 × 500 g jars

1.5 kg cherries

500 g apple, peeled and finely chopped

2 kg sugar

Juice of 2 lemons

- Place the ingredients in a pan. Heat slowly at first and then bring to a rolling boil, stirring constantly.

- After 4 minutes, check for a setting point.

- When set, pour into sterilised jars. Stores for 6 months.

GOOSEBERRIES

This is one of the most wonderful fruits on the planet. The combination of crunch and juice brings a sensation to the mouth that no other fruit can.

Growing for storage

The gooseberry is a hungry plant and, with all that sugar in the fruit, it is prone to fungal disease and another monster: the saw fly.

The key to good gooseberries is pruning. This should be done each winter and has two functions. First, it gives you a chance to open out the bush so the air can circulate, thus reducing fungal infections. Secondly, as the fruit forms on the previous year's growth, cut it back to two buds so you get fruit on these – it is easier to pick and the bush doesn't get overloaded.

Then feed in the spring and water regularly at the roots. Keep the saw fly at bay by covering with horticultural netting.

Harvesting

In June take half the crop and cook them. The second harvest will be bigger and produce better fruits that are for eating just as they are. Half of these can be preserved. Preserve only the best fruits and, once picked, put them straight into iced or cold water for 10 minutes, even if they are for jam or fool or freezing.

How to store

Gooseberries last only a few days on the shelf or in the fridge and, besides, one eats them too quickly anyway.

Gooseberries in Syrup

Makes about 3 × 500 g jars
500 ml water

500 g sugar

Approx. 1 kg gooseberries

- Boil the water and carefully add the sugar. Allow to cool and clear.

- Wash the gooseberries and place in sterile jars.

- Pour over the cool syrup to the jar's neck, making sure there are no air spaces. Seal and then boil the jars in a water bath for 15 minutes. Stores for about 6 months.

GRAPES

Until recently, the only grape variety available to the British gardener was Black Hamburg and it's still a good grower. However, today there are loads of varieties that give fantastic fruit and are easy care too.

Growing for storage

A polytunnel is probably the ideal place to grow grapes, especially in the cooler north of the country. Outdoors in the UK, however, many farmers are successfully growing commercial stock, which is mostly all converted into wine of ever increasing quality.

Indoor vines do well if they are fed with tomato fertiliser each month from a couple of weeks after they have burst into life in the spring, until the grapes are ready for picking. Since the bark is fibrous all kinds of pests overwinter, so scrape the bark away inside the tunnel or greenhouse.

The secret to good grapes is pruning, which is really beyond this book's scope. In essence, you run your grapes on vine wires that run the length of the tunnel. Prune out about 30% of the growth each year, leaving spurs for grapes to grow on.

Harvesting

The big problem with grapes is that, like gooseberries, they are full of sugar and unless you have good ventilation between the berries, penicillium fungi will infect the bunches. You can use scissors (some growers have special scissors just for the purpose) to thin out the berries so that the remaining grapes can grow unencumbered with a good air flow around them. The cardinal rule on harvesting is to cut off the piece of lateral they are growing from, so that you do not have to touch the grapes and either contaminate or damage them.

How to store

JUICE

Simply squash the juice out of them in a press and bottle it in sterile bottles. Pasteurise if you can or boil if not, and you will have some excellent juice for making into wine or simply drinking as juice. Bottled juice will store for 6 months.

RAISINS

If you have a lot of grapes, why not have a go at making raisins? You can use an oven, or just a tray in the greenhouse door if you can be sure of a couple of hot days. It takes about two days to dry them out or 10 hours using a dryer. First wash them under boiling water and then set to dry. They are ready when you squeeze them and no juice comes out.

Grape Jam

This is such a delicately flavoured jam everyone should make it, even if you have to go to the supermarket to buy the grapes. Seedless varieties are best. If not, it is possible to get the seeds out of the jam, it's just a little fiddly. You don't need water for this recipe, there is enough juice in the grape.

Makes 2–3 × 500 g jars

1 kg grapes

1 kg sugar

Juice of 2 lemons

- Add the ingredients to the pan and heat slowly until the grapes break and the sugar dissolves. Do this on a low heat – you don't want any caramelisation.

- Turn up the heat and bring to the boil. Boil for at least 4 minutes, checking for setting after this time. Repeat until you get a setting point.

- When setting point is reached, remove from the heat and cool for 2–3 minutes. Stir to evenly distribute the fruit and then pour into sterilised jars and seal.

Grape Wine

Obviously, if you grow grapes you need to make your own wine. Chateaux Peacock is a fruity little wine made from Black Hamburg that is so low in sugar most years I add 500 g of sugar when I am making it.

Makes 5 × 750 ml bottles

5 kg grapes

500 g plain white sugar

Juice of 1 lemon

1 teaspoon wine yeast

1 can wine starter

- Mash the grapes in a pan and decant the juice – you can use a fruit press for this job if you have one.

- Put the juice into a large pan and bring to the boil, adding the sugar.

- Allow to cool and then decant into a demijohn. Add the lemon juice and wine yeast. Seal with an airlock.

- Pour the wine starter (the appropriate colour for your grapes) into the wine juice demijohn.

- In a fortnight, rack off as per cider making: siphon the wine into a new sterile demijohn so that the lees are left behind and give it a good shake. Leave to stand for a week and rack again before bottling into sterile bottles. Stores for 1 year.

MELON

Growing for storage

Grown in polytunnels, melons are really easy so long as they have good rich soil with plenty of nutrients, especially phosphates, plenty of water, good drainage and heat. The truth of the matter is that in

Turkey escapee melons from gardens have become a serious problem in the wild where they grow as weeds, so they can't be hard to grow given the right conditions. Of course the problem we have here in the UK is with the heat; consequently you will have greater success indoors than outside.

If the melon is growing on the floor, then protect it with a mat or some straw. You can also protect them by holding them up in little nets. Maintain the watering every couple of days, at the roots, with correctly diluted tomato feed every second watering. If you overwater, the melons will taste watery.

Although they do like humid conditions, they will come to no harm in a drier polytunnel, and the general advice to water the path to increase humidity is often counterproductive in greenhouses where the increased chance of fungal infections would endanger other crops.

Harvesting

All melons except the Persian or Turkish types slip off the vine. You move the fruit about and it falls off! You can store these melons and they will tend to get sweeter.

The Persian types of melon remain fixed to the vine and you have to cut them off. Persian melons do tend to smell very aromatically when they are ripe, and you should use this as a guide for harvesting. They do not change much in storage at all.

How to store

They will last a couple of months on the shelf, ripening mostly. Allow plenty of air space around each melon for best storage. You don't have to keep them cool, just not too warm, which induces over-ripening.

FROZEN MELON BALLS

Melon doesn't freeze that well, but you can use a baller and make melon balls for freezing. Pop the balls in an ice cube tray, fill the space around them with melon juice and freeze. They make a great addition to drinks. Alternatively, fill the space with sugar syrup.

Rum

This is the traditional way to make rum.

1 large melon

Golden syrup, honey or molasses

- Slice the melon lengthwise. Scoop out the seeds and cut a groove to the top of the melon, so you can later insert a tube or a funnel leading to where the seeds were.

- Join the two halves together again and tape together with lots of tape. Pour, through the hole in the top, golden syrup, honey or molasses (it's quite hard to do this) into the melon until it is completely full. Plug the hole and also tape it up.

- Suspend the melon over a bucket and, after a fortnight, force a knitting needle into the melon from below. The melon will slowly drip a sweet, slightly alcoholic, melon-flavoured mixture – the original rum. Store in a sterilised bottle on the top shelf.

PEACHES

Growing for storage

Not all that easy to grow in temperate climates, peaches and nectarines suffer from cold, generally forcing their propagation indoors. However, the caveat is they need the greenhouse or polytunnel to be fairly cold in the winter, which is not always what you want when growing other crops. Some people grow peaches in pots so they can be moved around – brought out of the tunnel in the summer, taken indoors in winter. You have to be fastidious about replenishing the compost!

They can suffer from peach leaf curl, something that destroys the crop. If your leaves are reddening and curling, remove them ASAP, and the flowers and any fruit. Bit drastic that. You can also spray with Bordeaux Mixture which, although no longer an organic solution, has been used for centuries without any problems for human health.

Harvesting

When the fruit come off the branch easily they are mostly ripe. They will soften over time, but not improve in flavour at all. If you squeeze them they will give way a little and also smell very strongly of peach.

How to store

They will last about a week, at best, in a plastic bag in the fridge.

Peaches in Syrup

I have to say, I am only happy if the peaches have been peeled.

Makes approx. 2 large Kilner jars

500 g sugar

1 litre water

1 kg peaches

- Boil the water and carefully add the sugar. Allow to cool and clear.

- Wash the peaches, peel, de-stone and slice into quarters.

- Pack into sterilised jars. Fill to the neck with the sugar syrup, making sure there are no air spaces. Seal and boil for 15 minutes in a water bath. Stores for 6 months.

Peach Chutney

Makes 4–6 × 500 g jars

2 kg peaches, chopped

4 cloves garlic, finely chopped

300 g raisins

500 g sugar

600 ml white vinegar

25 g fresh root ginger, finely chopped

1 level teaspoon mixed spice

2 teaspoons salt

- Place the peaches in a pan with the other ingredients.

- Heat gently, stirring until all the sugar has dissolved.

- Raise the heat and bring to the boil. Then simmer for 1 ½ hours until thick and smooth.

- Pot in sterilised jars and seal immediately. Label when cool. Stores for a year.

PEARS

Growing for storage

Generally, I treat pears like apples, to which they are closely related. Usually you need a pollinator so more than one tree is needed, and of course you have a wide variety to choose from. Plant them well, digging deeply and incorporating plenty of rich organic matter. Keep them well watered and well fed.

Harvesting

Do not harvest fruit for three years. I remove the fruit in the first and second years, and from there I take only a couple, increasing this quantity over the years. They will fruit for 30 years, so a good start is all important.

Generally the ripening process starts when the pears begin to change colour. Take them as soon as this happens. They should twist off, but don't pull them off. If all else fails, cut them, but don't rip them off the tree.

How to store

You can store pears like apples: singly, wrapped in tissue or kitchen paper. The temperature should be around 15°C and the fruit should last for about two months. You can increase this by a month if you keep them in the fridge.

Pears in Syrup

Makes approx. 3 × 500 g jars

1 litre water

500 g sugar

1 kg pears

Grapes and apple (optional)

- Boil the water and carefully add the sugar. Allow to cool and clear.

- Wash the pears and quarter them, removing the pips. I like long, thin strips, but you might prefer chunks. You can also add some grapes and apple – make up your own fruit cocktail!

- Pack the fruit into sterilised jars and cover with syrup to the neck, making sure there are no air spaces. Seal and boil the jars for 15 minutes in a waterbath. Stores for 6 months.

Pear Jam

When you cook pears they fall easily, well especially when you make jam. I tend to use pears with other fruits, pear and raspberry being my favourite. On the whole, use 75% pear and 25% other fruit.

Makes 3–4 × 500 g jars

2 kg pears, peeled and cored

2 kg sugar

Juice of 2 lemons

- Add the pears to a large pan and heat slowly on a low heat until the pears fall and the sugar dissolves.

- Turn up the heat and bring to the boil. Boil for at least 4 minutes, stirring all the time, checking for setting after this time. Repeat until you get a setting point.

- Then remove from the heat and cool for 2–3 minutes. Stir to evenly distribute the fruit and then pour into sterilised jars and seal. Stores for 6 months.

JUICE

Pears are best juiced using a steam juicer – something I have realised since obtaining one. Otherwise, it works quite well if you chop or crush your pears and then press the juice out. Pear juice is best combined with the juice of 1 lemon per litre. It freezes well and can be kept for 6 months.

Having bottled the pear juice, you can pasteurise it to keep it for a month, or boiled it will last all year.

Pear Wine

Makes approx. 5 × 750 ml bottles

4.5 litres pear juice – I always pasteurise my juice before making wine.

250 g sugar

1 teaspoon white wine yeast

- Transfer the juice to a demijohn and add the sugar and white wine yeast. Then fit an airlock. Be sure everything is sterile.

- Leave to ferment, which should take about two weeks. Be sure to stand the demijohn in a bowl, as this wine often blows through the airlock.

- The answer is to rack once the bubbles have stopped and then rack again about a week later. If you can, stand the settling wine on a stone floor – it settles better than on vibrating floorboards.

- The wine will be ready after a month and will store forever.

DRYING

Wash, core and slice your pears into 5 mm slices. I like long thin strips. Sprinkle both sides with 1 tablespoon of caster sugar per kilo of fruit and then place in the desiccator for about 8 hours on its lowest setting. Then store in a jar on top of some dry rice and they will last all year – though I cannot really vouch for that because I have usually consumed them well before. They are ready when they are rubbery.

PLUMS

Once you have a plum tree settled and productive, there is no more wonderful plant for a summer's joy. Plums are so full of sweetness and vitamins, they should be grown in every garden.

Growing for storage

Establishing a plum tree is much the same as an apple tree. Incorporate lots of well rotted manure and firm well, supporting your plant with a stake. The secret to establishing them for great fruit is to water every day in the first few months of life. Copious water until the plant is established eases the roots into their new position, and gives the plant a chance to grow a whole load of root hairs.

Harvesting

It takes a good few years for your tree to produce plums at all. Feed every spring, and around the fourth year you will get fruit. They are best taken, as with most fruit, when they are easily twisted off the tree. Windfalls are best cooked, as they will be damaged by their fall.

How to store

Plums do not keep well fresh, little more than a week or so, and consequently they have to be stored frozen or as jam, etc.

FREEZING

They fall apart when frozen so the best way is to stew them first, because they will only need stewing later on! I keep the skins on, slice and stone them. Add a little water (around 50 ml) to a pan and add your plums. Bring them to the boil and cook for a few minutes. Allow to cool. Then transfer to a freezer bag or sterilised ice cream tub for freezing. Stores for 6 months.

PRUNES

It's not that difficult to make prunes, especially if you have a clear few days of good summertime heat. Wash, slice and stone your plums and lay them on a tray. I actually use a Beekeeping Varroa Mesh floor, which allows drying on all sides. You can stand them in the polytunnel door to keep any rain off, but cover with a net to keep the insects away. Drying outdoors takes about three days.

Alternatively, you can use the lowest setting on your electric dryer and drying will take about 12 hours.

Plums in Syrup

Makes approx. 3 × 500 g jars
1 litre water

500 g sugar

1 kg plums

- Boil the water and carefully add the sugar. Allow to cool and clear.

- Wash and stone the plums.

- Pack into sterilised jars and cover with syrup to the neck, making sure there are no air spaces. Seal and boil the jars for 15 minutes in a waterbath. Stores for 4–6 months.

Plum Jam

I always bulk this out with apples as they add a firmness to the jam.

Makes 6–8 × 500 g jars

1.5 kg plums

500 g apples, peeled and cubed

600 ml water

Juice of 2 lemons

1.5 kg sugar

- Add the fruit to a pan with the water and lemon juice.

- Bring to the boil, then simmer until the fruit is soft and pulpy.

- Add the sugar and stir over a medium heat until dissolved.

- Bring to the boil and boil for 5 minutes. Then test for setting.

- Allow to stand for 3 minutes then ladle into sterilised jars. Secure the lids well. Label and date the jars when cool. Stores for 6 months.

RASPBERRIES

The following information applies to all berries of this type: logan, tay, as well as the myriad varieties to be found in the shops. This Rubus species is not only inter-fertile, they frequently double their chromosome numbers and come up with new varieties, all by themselves.

Growing for storage

The best way to grow raspberries for storage is to feed them in the spring. Give them a mulch of good quality compost – don't let the compost actually touch the canes, but spread it around them. If you can incorporate 50% very well rotted manure in this compost all the better.

When the cane has finished fruiting, leave it to grow until the autumn, when you should cut it down as low as possible. They are grown on a wood and wire fence, and you should tie in the new canes that will fruit next year at the same time as cutting out the old cane.

They are thirsty plants, so make sure they never dry out.

How to store

Being largely water, these berries do not last well on the shelf or in the fridge, but they are excellent juiced.

JUICING

Boil a muslin bag to sterilise it and place your washed berries inside. You can then press the juice out of the berries for use. It is great mixed 50:50 with apple juice, or used alone. I always add 250 g of sugar per litre of juice and then pasteurise. The juice then keeps bottled for up to 2 months.

FREEZING

There are two ways to freeze raspberries and both methods store for 6–8 months. The more successful is to boil them down with a tablespoon of sugar and spoon into a freezer bag when cool. This is wonderful on ice cream!

The other way is to place your washed raspberries on a tray, sprinkle liberally with sugar and freeze. When frozen they can be gathered together and stored in the same container. You can freeze without the sugar, but it is not quite as successful.

Raspberry Jam

Note the lack of complex ingredients – jam should be as natural as possible.

Makes about 8 jars

  2 kg raspberries

  2 kg sugar

  Juice of 1 lemon

  • Place the fruit in a pan and simmer gently for about 10 minutes until the fruit is soft and the juice is running.

- Remove from the heat and add the lemon juice.

- Stir in the sugar and replace on a low heat. Stir gently but constantly until the sugar is completely dissolved.

- Turn up the heat and bring to the boil. Boil for 4 minutes, checking for setting after this time.

- When setting point is reached, remove from the heat and cool for 2 or 3 minutes. Stir to distribute the fruit evenly, then pour into sterilised pots and seal. This jam should keep for 6–9 months unopened.

Raspberry Wine

This product usually tastes like real moonshine – so to enhance the flavour, add a couple of tablespoons of homemade raspberry jam. Dissolve the jam over a gentle heat and then try to get the seeds out. I say homemade jam because the shop-bought stuff might contain preservatives that will deter the yeast.

Raspberries – as many as you can get!

500 sugar per 4.5 litre juice

1 teaspoon claret wine yeast per 4.5 litre juice

- First pasteurise your raspberry juice and add 500 g sugar for each 4.5 litre of juice. Dissolve the sugar and pour into a sterile demijohn. I usually collect about 45 litres of juice and have a series of demijohns on the go in the late summer.

- Add the wine yeast and let it ferment until the bubbles stop. Rack, rack and rack again.

- The wine will take about 6 months to taste really nice, and then have a liver doctor on hand just in case. It will keep for about a year.

167

RHUBARB

Growing for storage

Where would we be without rhubarb? It is the best vege-fruit in the world. The best thing about rhubarb is that it comes on line early in the year, and almost without interference from you. The best time to plant is in the winter, when you dig in plenty of well rotted manure and compost.

Rhubarb will grow more or less anywhere as long as the soil is free draining, but not dry by any means. If the soil becomes waterlogged you are likely to lose the crown to fungal infection. Try to place it where there will be no shade.

Some people prefer to blanch (which in this case means keep in the dark) to reduce the flavour and colour. You can also get self-blanching varieties.

Harvesting

Harvest in late spring to early summer when the stalks are about 3 cm thick. They pull easily from the crown. You should not eat the leaves, as they are rich in oxalic acid, which can be harmful. As the summer turns, by the end of June, you should not be using the stalks either, because of the increasing oxalic acid content.

How to store

Rhubarb lasts about 2 weeks on the shelf, wrapped in cling film which keeps the moisture in. Storage is increased by a week in the fridge.

FREEZING

Simply chop the rhubarb into pieces no bigger than 3 cm. Wash and freeze on a tray and when frozen collect into a freezer bag for storage between 6 and 8 months. You can freeze stewed rhubarb, with a tablespoon of sugar to a kilo of rhubarb, in the same way, which is wonderful for pouring over ice cream and so on. Incorporate frozen rhubarb into recipes without thawing – the pieces become mushy as the freezing water bursts the cells.

STRAWBERRIES

Growing for storage

Strawberries get their name from the fact that they were traditionally grown with straw under the leaves so that the fruit would not rest on the soil and become spoiled. The straw also makes for an excellent deterrent against slugs and snails. They are grown in pots for the same reason, with the fruit dangling over the side or resting on pebbles. You can also use strawberry mats, which are a bit like plastic beer mats, only bigger, but I have found they are no good for keeping slugs at bay.

Strawberries are at their best when they are grown in a sunny position with good drainage. They are very hardy plants and will easily last out a winter with little problems, but new growth in the spring seems to be more susceptible to frost. Young flowers seem to be particularly vulnerable, and the best way of countering this is to plant them in the highest point of the garden. Another alternative, especially where slug and snail avoidance is concerned, is to grow them in a hanging basket.

Harvesting

Take strawberries when they are big enough to eat. Try to avoid eating them as soon as they come off the plant!

How to store

They last for not more than a few weeks on the shelf or in the fridge.

Whole Fruit Strawberry Jam

Makes 10–12 × 500 g jars
  1 Bramley apple

2 kg strawberries, remove the calyx (green bits) and rinse

2 kg sugar

Juice of 1 lemon

- For added pectin, peel the apple and chop into small pieces. Place everything (including the peel) in a muslin bag.

- Add the strawberries, apple and sugar to a pan and heat together over a low heat. Try to stir only to keep the mixture moving – too much and you will break up the fruit.

- Add the lemon juice. The sugar will gradually dissolve and the juices of the fruit should start to run.

- When the sugar is fully dissolved bring quickly to the boil.

- Test for setting after 4 minutes of continuous boiling. It should have reached its optimum setting point, but it won't be a firm set.

- Remove from the heat, stir gently and pot in sterilised jars. Stores for 6 months.

Strawberries in Syrup

Makes 4–5 × 500 ml jars

1 litre water

500 g sugar

2 kg small strawberries

- Boil the water and carefully add the sugar. Allow to cool and clear.

- Remove the calyx (the leafy bits at the end) from the strawberries and wash.

- Pack into sterilised jars and cover with syrup to the neck, making sure there are no air spaces. Seal and boil the jars for 20 minutes in a waterbath. Stores for 3 months.

Strawberry Purée

Strawberries

Dash of balsamic vinegar

1 teaspoon sugar per 250 g strawberries

- Smash the strawberries in your food processor, along with balsamic and sugar.
- Freeze the purée in individual containers, enough for a single use, so you don't have to bash your way through frozen purée. Stores for 6 months.

CHAPTER 8

GROWING AND PRESERVING HERBS FOR THE KITCHEN

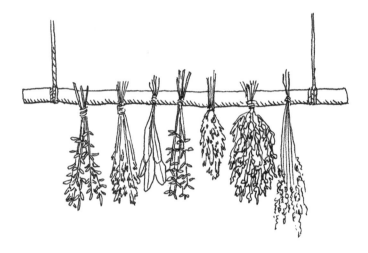

People grow herbs for the pot and for the medicine cabinet and often a preserved herb is better than fresh. This chapter looks at how to preserve your herbs all the year round.

Herbs are interesting in that they are often more pungent dried than fresh, as though some of the active ingredients become concentrated. It is easy to grow herbs, but not so easy to grow them well, and whereas some of them are best fresh, preserving herbs is an important part of this craft.

Herb growing is a skill, there is no doubt about it. The usual advice about feeding regimes and watering holds true for many plants, but there are others where overfeeding and moderate watering will kill them. So you need to learn about the plants you are growing and this takes time.

GROWING IN POTS

There are many reasons why I grow my herbs in pots. One reason is so that I can have them close to my kitchen door, which means that I can pick them whenever I need them and they are always to hand. Another advantage is that they can be moved around the garden or patio easily so you can find the perfect place for them to thrive. For some very invasive herbs, such as mint and lemon balm, growing in pots keeps them in one place and stops them from growing everywhere.

They are relatively easy to grow in pots, although watering is very important, especially for herbs grown in terracotta pots, as they dry out much quicker than when they are grown in plastic or other material. If you do like growing them in terracotta, line the bottom of the pot with some plastic liner and allow it to come one-third up the side of the pot. It will still require some drainage holes at the bottom, however.

The main reason for growing in pots is to be able to provide the perfect environment for that particular herb, and still have it conveniently placed next to another plant that might need a totally different regime.

GROWING HERBS FOR FRESHNESS

It is important that herbs are well hydrated just prior to picking – perhaps the day before they should be watered and allowed to rest. This gives the chemistry of the plant enough time to create the optimum amount of chemical ingredients for use in the kitchen or medicine cabinet. Chives, in particular, take on a slightly bitter flavour when allowed to dry out.

Once picked, your leaves or stems or whatever portion you are going to use start to deteriorate, so use them quickly. I used to wonder why people are so insistent on growing herbs by the kitchen door, as though a little way into the garden was too far to walk. But in my latter years I have discovered it is too far!

SOWING

In spring, sow the seeds in new compost. A soil-based compost is the best, as it contains a high level of nutrients on which the herbs thrive. I plant them in individual modules, planting 2–3 seeds in each one. Prick out the least vigorous plant as they begin to grow. Once they are sown, cover with polythene or a plastic lid and keep warm. I leave them in the kitchen or the conservatory. Keep them well watered using a spray rather than a watering can, as this is far more gentle on the tiny plants.

AVOID FUNGAL INFECTIONS

As we have already said with other plants, the fungus that causes damping off can infect plants without really causing immediate damage, but may go on to produce damage later. So, although you may not see signs of damping off, your plants could still be infected. For this reason, make sure you remove the lid from your seedlings at least daily to reduce humidity and the likelihood of infection.

AFTER GERMINATION

As the plants grow they should be very carefully transferred to a larger pot to develop their root system. Once again, pot in soil-based compost and use an 8 cm pot. They can be kept in the greenhouse but protect from any late frosts – I tend to use a little fleece if necessary.

LOOKING AFTER YOUR HERBS

In the summer, always keep your herbs moist but if there is a heavy downpour, shelter them as they don't like to be drowned. Feed weekly, preferably with a seaweed-based liquid feed. Pick the leaves regularly, but be careful not to overpick as they will need some of their leaves to keep growing.

It is a good idea to have a fortnight run for herbs. Something like mint, that we use regularly, occupies 14 pots around our garden. It is fairly rare you have to pick mint every day; in my experience I need it a couple of times a week. Having 14 (that's where the fortnight comes in) allows you to pick from one on day 1, and from the next pot at a later date and so on. This way you never devastate your plants all in one go, making it difficult for them to recover.

When you harvest, always feed the plant as soon as possible afterwards. The boost of nutrients aids the healing process.

Pick off any flowers that appear in early summer, as this will make for more vigorous plant growth, but later keep some to dry, or wait until seeds appear and then store some for the following season.

I usually keep some pots of herbs for flowers and some for seeds. If you are going to collect flowers, do so before they are pollinated, then they will be richer in fragrance, flavour or colour. Plants being kept for their leaves, root or stem, should not be allowed to flower at all.

When the weather starts to get colder, place the pots in a sheltered spot or back in the greenhouse, or bring them into the kitchen so you can continue to use them in your cooking easily.

If your sage or thyme is getting too big for its pot, just transplant into a slightly larger pot and top up with some extra fresh compost. This is best done on a bright, sunny day. Again, water well and allow to stand in a sunny spot to recover.

DRYING AND PRESERVING HERBS

Perhaps the easiest way to preserve herbs is in an ice cube tray. This goes for almost every herb, and I learned the trick on my son's wedding day. It was a rare affair, and I drank rather too many Pimm's – in fact, looking back I probably drank everyone's Pimm's and quite probably made a fool of myself. Inside an ice cube in each drink was a borage flower, one of the herbs in the drink.

This, I thought, was an excellent way of storing any fresh herb. Just put enough for one measure in each cell of an ice cube tray and then fill with water. It doesn't matter if you have to fold the leaves, or even chop or crush them, any aromatic molecules will simply go into the water as you top up each cell. Freeze the tray and if you need an easy portion of something, simply pop it into the pan! Herbs prepared in this way are usually used for stews and soups, and so the addition of a few millilitres of water doesn't matter. This works particularly well with garlic. Frozen herbs will store for a year.

WHEN TO DRY HERBS

Taking fresh herbs at any time as they grow is one thing. Cutting herbs for drying is another. The cooler the weather, the more herbs deteriorate and this means that the middle of the summer, towards August, is probably the best time. Try to pick before flowering and make sure the weather is not too hot.

Before you pick be sure the plants are fully hydrated and have plenty of nutrients. Having collected, be sure there is no dew on the leaves, which can cause rot later on – simply leave them to dry on tissue paper.

Traditionally, herbs are simply tied at their base (the stems) with butcher's twine or something similar. Don't use cotton or wool, which will either cut or bruise the stems. Hang them upside down in a cool, dry, draughty place – not in the kitchen, where excessive heat will dry them too quickly or aromas will spoil them.

It takes about three weeks to dry herbs, after which they should be placed in a sealable box, loosely wrapped in kitchen paper. Try to keep them separate: don't mix mint or sage and so on with other herbs, you only want to mix them at the last minute. Similarly, only rub or cut or grind dried herbs at the time you need to use them for a dish. Don't store them ready rubbed as it were, as they lose their aromas too quickly.

USING A DEHYDRATOR

These tools are great fun to use and enable you to dry herbs at any time of the year, rather than having to wait for summer.

When drying herbs, make sure you are able to use the lowest setting and even then be careful not to overdo the drying. It is a good idea to pack the dryer with quite a lot of one type of herb to avoid contamination. Pack the trays as fully as you can (this will help avoid overdrying the herbs) and on the lowest setting they should be done in 24 hours.

MAKING A BOUQUET GARNI

For most purposes a bouquet garni is a collection of herbs, dried and tied, to flavour a casserole of some kind. It usually comprises a bay leaf, one or two springs of thyme and three or four sprigs of parsley. You can also use these herbs fresh. They are frequently tied in muslin and placed in the food to be removed at the end of the cooking period.

Vary your basic bouquet for different dishes: for pork, add some sage and rosemary, and for poultry, you could add a sprig of rosemary.

Dried bouquets are easy to make. Simply sew some muslin bags into which you place the fresh herbs. Keep them in an airy place for a week and then store in a box lined with kitchen paper.

Herbes de Provence

This is a bouqet garni with strongly flavoured herbs – particularly the basil cousin, oregano (as is marjoram). The basic recipe is really a matter of taste and, to be honest, we miss out the oregano because we don't like the flavour.

5 tablespoons dried thyme

3 tablespoons dried savory

2 tablespoons dried marjoram

5 tablespoons dried rosemary leaves

1 ½ tablespoons dried lavender flowers

Store the lot in a sealed container and mix well. Use 1 tablespoon placed in a muslin bag for cooking.

HERB VINEGARS

You have to remember that a few herbs in vinegar is probably one of the best tonics you can have for a pick me up. Just pour a little into your bath. Try a mix of sage, lavender and mint in cider vinegar for a refreshing, gently rejuvenating bath. Use half a cup in the bathwater.

Of course, there are vinegars for eating to be made too. Try mint alone or thyme, but most of your favourite herbs can be vinegared. Simply crush the leaves by rubbing them gently in your hand. Add them to a sterile bottle (sterilise the lid also) and then cover with cider vinegar. You should have about 24–50 g herbs in 500 ml of vinegar.

INDEX